ATLAS OF DESCRIPTIVE EMBRYOLOGY

SIXTH EDITION

Gary C. Schoenwolf, Ph.D.

Professor, Department of Neurobiology and Anatomy
Director, Children's Health Research Center
University of Utah School of Medicine

Willis W. Mathews, Ph.D.

Formerly Professor, Department of Biological Sciences
Wayne State University

Editor-in-Chief: *Sheri Snavely*
Project Manager: *Karen Horton*
Executive Managing Editor: *Kathleen Schiapavelli*
Assistant Managing Editor: *Dinah Thong*
Production Editor: *Veronica Malone*
Manufacturing Manager: *Trudy Pisciotti*
Manufacturing Buyer: *Ilene Kahn*
Cover Manager: *Jayne Conte*
Cover Designer: *Bruce Kenselaar*

© 2003 by Pearson Education, Inc.
Pearson Education, Inc.
Upper Saddle River, NJ 07458

The author and publisher of this book have used their best efforts in prepar-
ing this book. These efforts include the development, research, and testing of
the theories and programs to determine their effectiveness. The author and
publisher make no warranty of any kind, expressed or implied, with regard to
these programs or the documentation contained in this book. The author and
publisher shall not be liable in any event for incidental or consequential dam-
ages in connection with, or arising out of, the furnishing, performance, or use
of these programs.

Printed in the United States of America

10 9 8 7 6 5 4 3 2 1

ISBN 0-13-090958-0

Pearson Education Ltd., *London*
Pearson Education Australia Pty. Ltd., *Sydney*
Pearson Education Singapore, Pte. Ltd.
Pearson Education North Asia, Ltd., *Hong Kong*
Pearson Education Canada, Inc., *Toronto*
Pearson Educatíon de Mexico, S.A. de C.V.
Pearson Education—Japan, *Tokyo*
Pearson Education Malaysia, Pte. Ltd.
Pearson Education, *Upper Saddle River, New Jersey*

Preface

The sixth edition of Atlas of Descriptive Embryology is my second revision of this popular atlas. My goal in this revision, as in my last revision, was to continue to meet the need, as stated by Professor Mathews in the Preface of the First Edition, for an atlas consisting of detailed, accurate pictures of a wide range of standard laboratory materials, which are fully labeled. To this end, I have made principally four revisions in the sixth edition. First, a new chapter (Chapter 3) has been added on worm development, using *Caenorhabditis elegans* as a model organism. This organism serves as an important model system for understanding mechanisms of development at the genetic, molecular and cellular levels. Its complete cell lineage, which is invariant, has been deciphered and its entire genome has been sequenced. Additionally, many interesting mutants are available. Second, all figures have been renumbered according to the chapter in which they appear and their order of appearance within that chapter. Thus, for example, Figure 8.5 is the fifth figure appearing in Chapter 8. Third, leader lines on all figures have been modified to increase their visibility. Finally, the Glossary, Synopsis of Development and Index has been revised to increase its usefulness and accuracy. Because my overriding goal in revision is to make succeeding editions better than previous editions and, in particular, to make each new edition more useful to students and instructors, I invite your comments (Schoenwolf@hsc.utah.edu).

On a personal note, it remains a pleasure to work with the superb photomicrographs produced by Professor Mathews during the course of the first four editions of this atlas. The embryo is a beautiful and wonderful organism; it deserves nothing less than to be accurately and artistically portrayed. I believe this new edition does exactly that. I hope that you will concur and that you will enjoy following "the way of the embryo."

For this edition, special thanks is due to Dr. Susan Mango, Hunstman Cancer Institute, University of Utah. She provided beautiful embryos, unending enthusiasm and scientific expertise that made Chapter 3 possible.

Contents

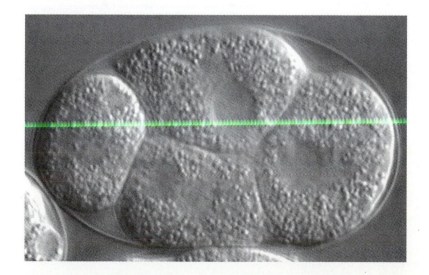

Chapter 1

Gametogenesis

Mammals: **Figures 1.1-1.9**
Insects: **Figures 1.10-1.15**

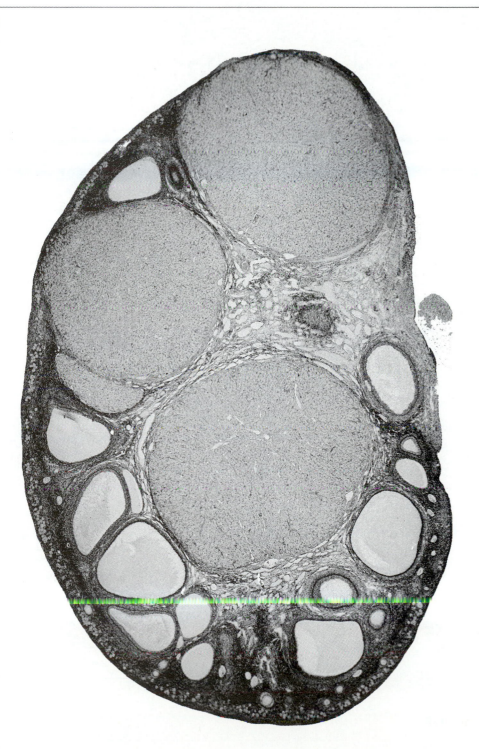

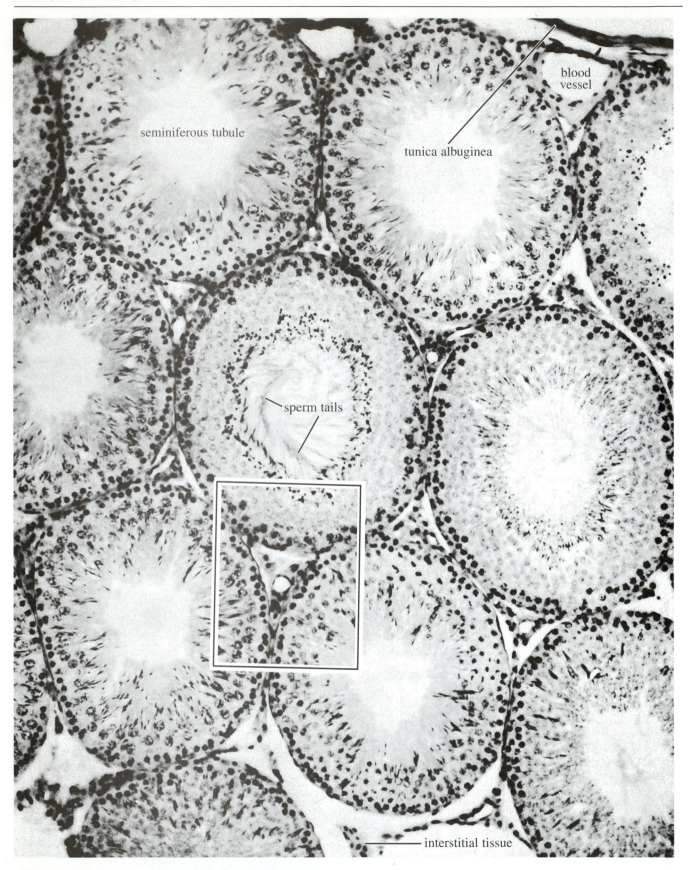

Figure 1.1

Mature rat testis, section (200X). The boxed area indicates the area shown in Figure 1.2.

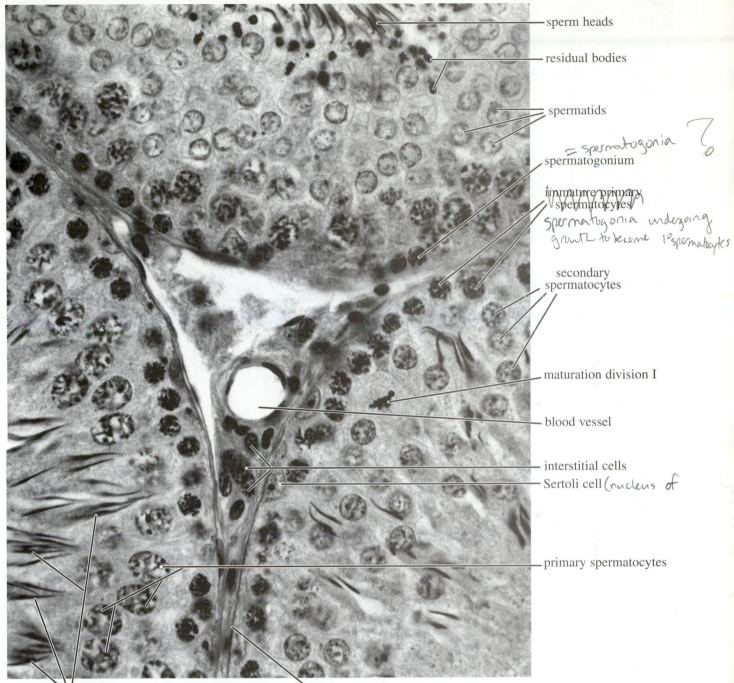

sperm heads

residual bodies

spermatids

spermatogonium   = spermatogonia ?

immature primary spermatocytes   Spermatogonia undergoing growth to become 1° spermatocytes

secondary spermatocytes

maturation division I

blood vessel

interstitial cells

Sertoli cell (nucleus of

primary spermatocytes

sperm heads   basement membrane of seminiferous tubule

Figure 1.2

Mature rat testis, section (800X). Leader lines indicate cell nuclei. Outlines of surrounding cytoplasm are indistinct.

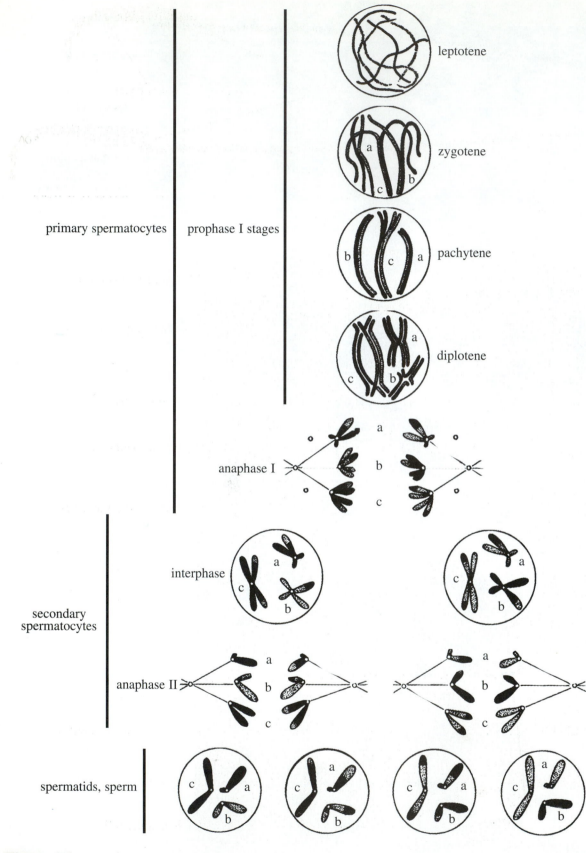

Figure 1.3

General diagram of meiosis, illustrating the union, separation and distribution of the chromosomes.

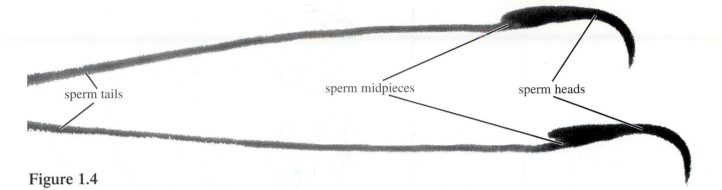

Figure 1.4

Rat sperm smear (250X).

sperm tails

sperm midpieces

sperm heads

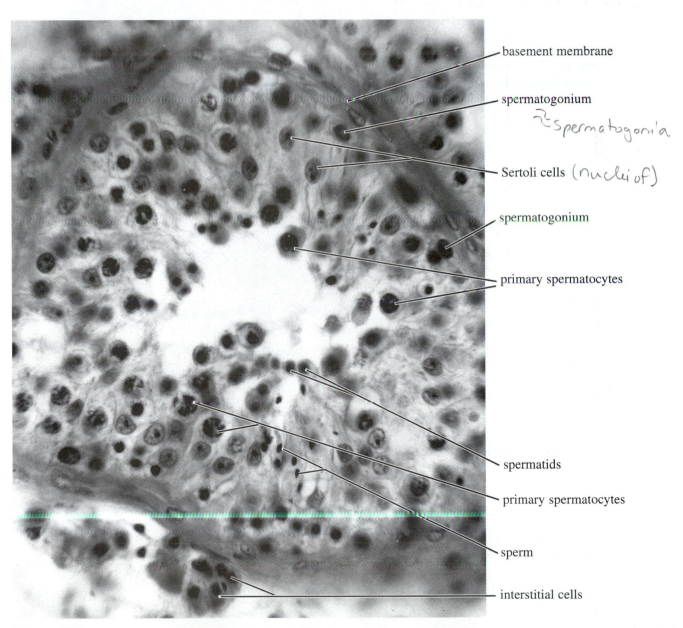

basement membrane

spermatogonium

≈ spermatogonia

Sertoli cells (nuclei of)

spermatogonium

primary spermatocytes

spermatids

primary spermatocytes

sperm

interstitial cells

Figure 1.5

Human testis, section of seminiferous tubule (675X).

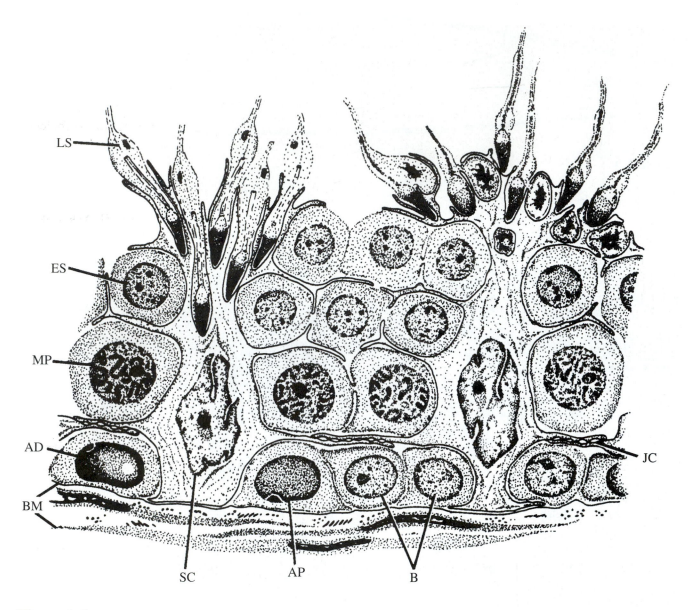

Figure 1.6

Diagram of a portion of a seminiferous tubule showing a group of germ cells and their relationship to the Sertoli cells that extend the entire width of the tubule wall. MP, middle pachytene spermatocytes; ES, early spermatids; LS, late spermatids; AD, type A dark spermatogonium; AP, type A pale spermatogonium; B, type B spermatogonium; BM, basement membrane; SC, Sertoli cell; JC, junctional complex.

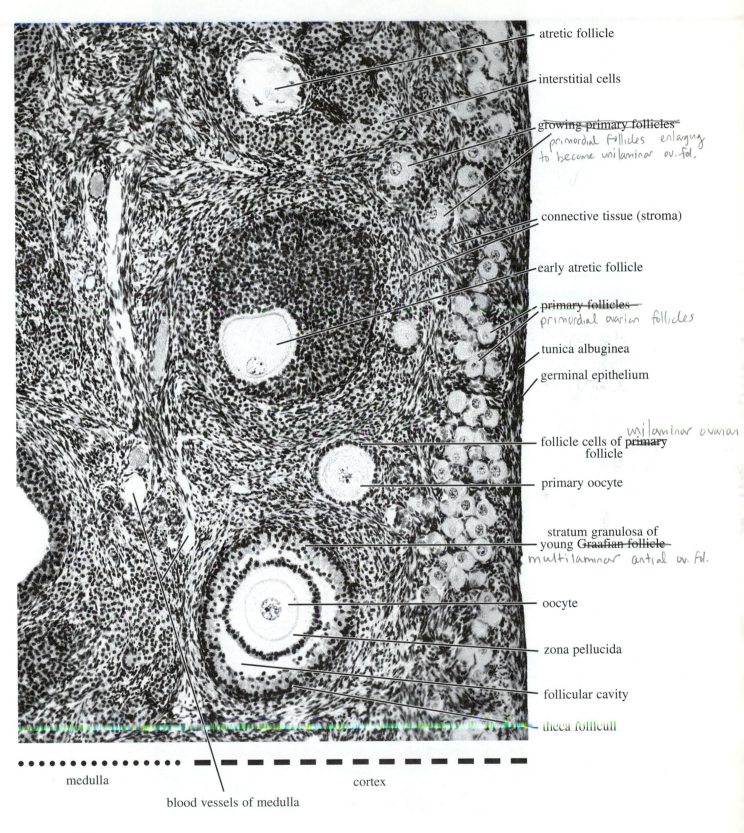

atretic follicle

interstitial cells

growing primary follicles
primordial follicles enlarging to become unilaminar ov. fol.

connective tissue (stroma)

early atretic follicle

primary follicles
primordial ovarian follicles

tunica albuginea

germinal epithelium

follicle cells of primary follicle
unilaminar ovarian

primary oocyte

stratum granulosa of young Graafian follicle
multilaminar antral ov. fol.

oocyte

zona pellucida

follicular cavity

theca folliculi

medulla

cortex

blood vessels of medulla

Figure 1.7

Mature cat ovary, section through cortex (150X).

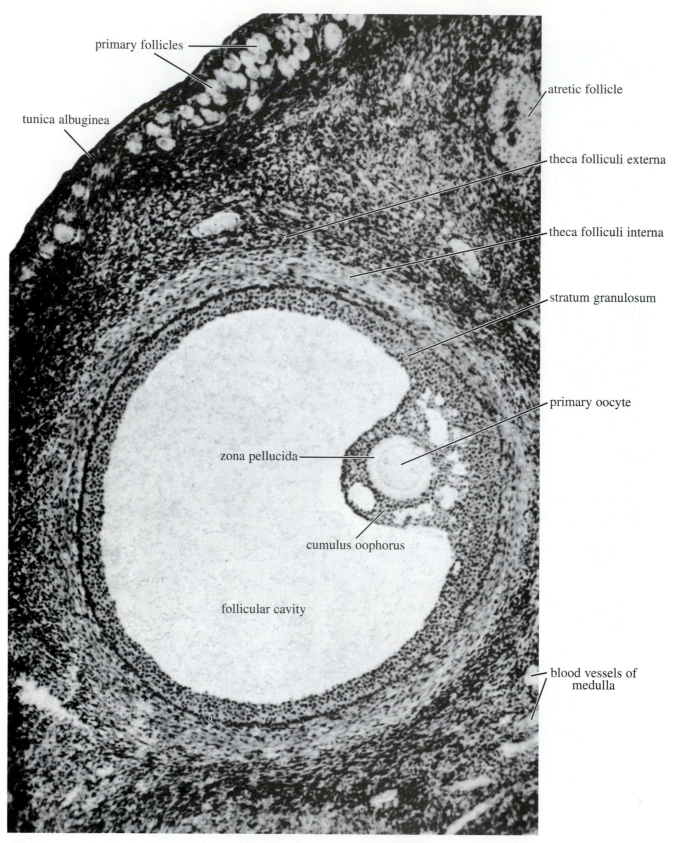

primary follicles

tunica albuginea

atretic follicle

theca folliculi externa

theca folliculi interna

stratum granulosum

primary oocyte

zona pellucida

cumulus oophorus

follicular cavity

blood vessels of medulla

Figure 1.8

Mature cat ovary, section through large Graafian follicle (115X).

multilaminar
ovarian follicle

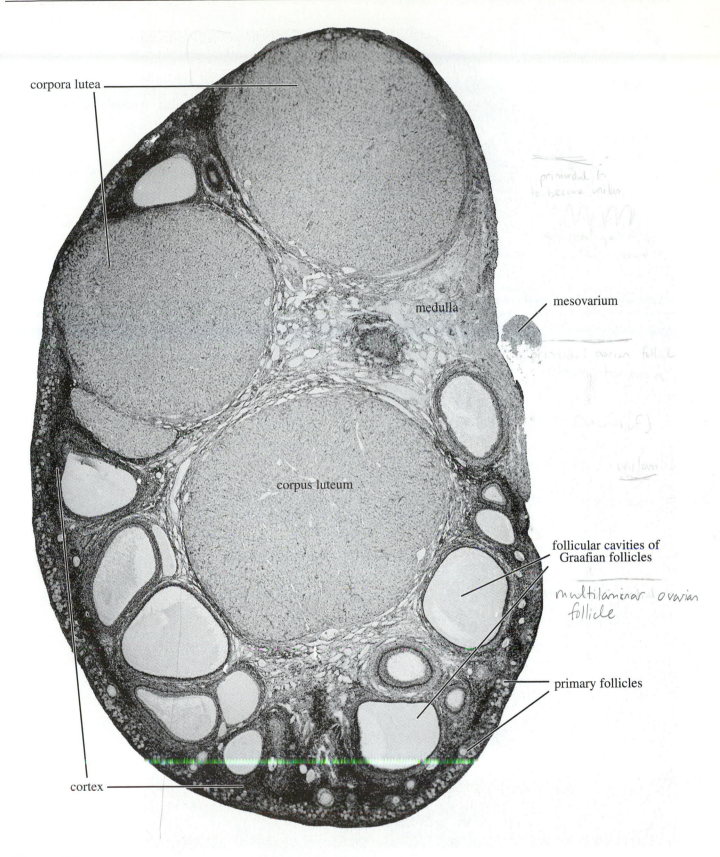

corpora lutea

medulla

mesovarium

corpus luteum

follicular cavities of
Graafian follicles

primary follicles

cortex

Figure 1.9

Ovary of pregnant cat, section through corpora lutea (35X).

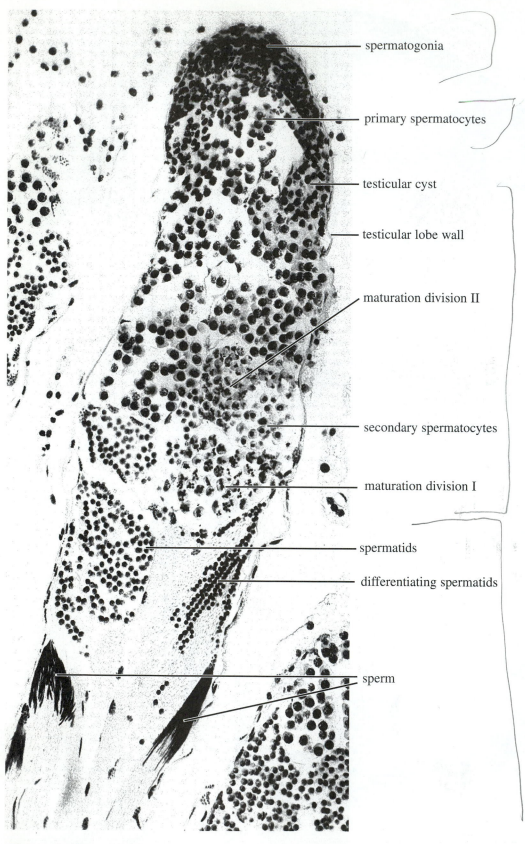

spermatogonia

primary spermatocytes

testicular cyst

testicular lobe wall

maturation division II

secondary spermatocytes

maturation division I

spermatids

differentiating spermatids

sperm

Figure 1.10

 Grasshopper testis, longitudinal section of testicular lobe, Feulgen stain for DNA (190X).

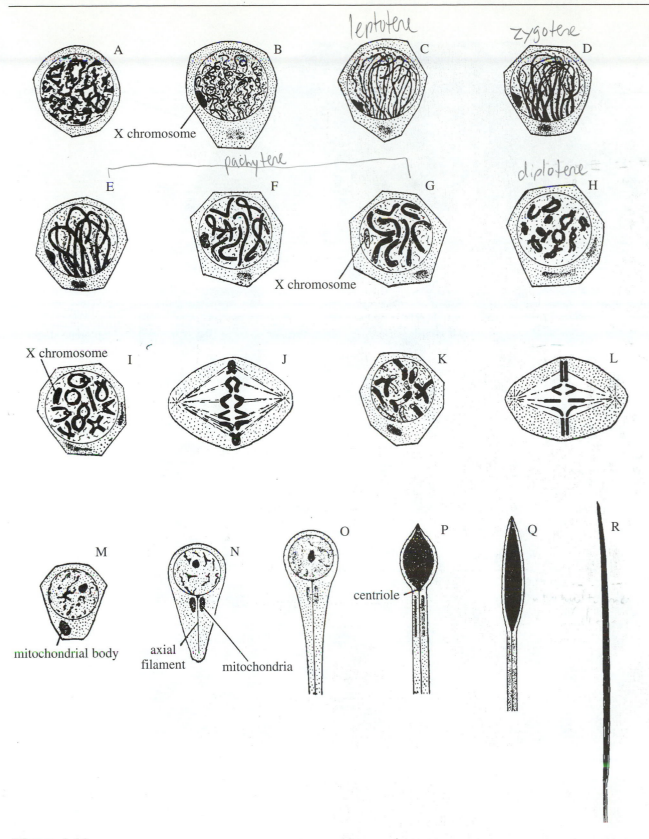

Figure 1.11

Spermatogenesis in the locust, *Rhomaleum tricopterum*. a, spermatogonium; b-j, primary spermatocytes; b, unraveling chromosomes; c, leptotene; d, zygotene; e-g, pachytene; h, diplotene; i, diakinesis; j, metaphase; k, l, secondary spermatocytes; k, prophase second meiosis; l, metaphase second meiosis; m, spermatid; n-r spermiogenesis.

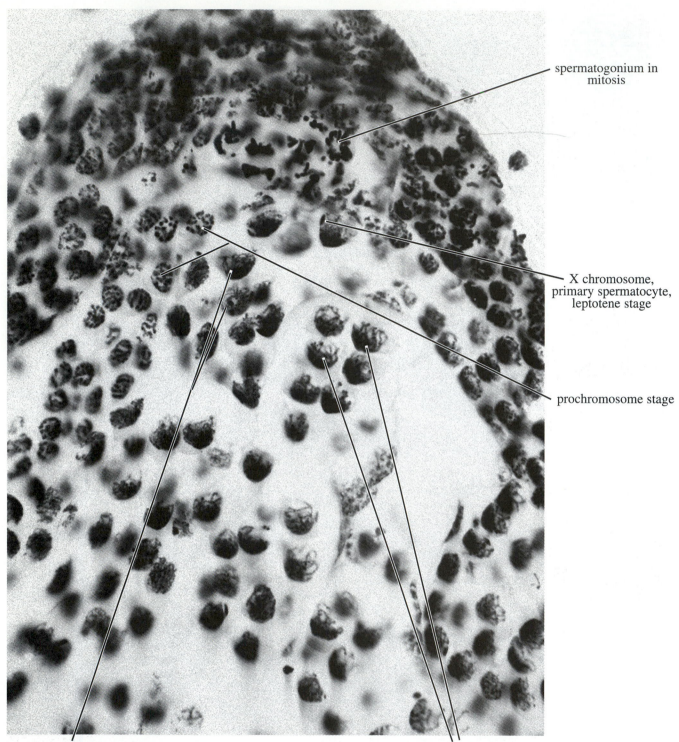

spermatogonium in
mitosis

X chromosome,
primary spermatocyte,
leptotene stage

prochromosome stage

leptotene stage, primary spermatocyte

early pachytene stage, primary spermatocyte

Figure 1.12

Grasshopper testis, longitudinal section showing area from Figure 1.10, Feulgen stain for DNA (1320X).

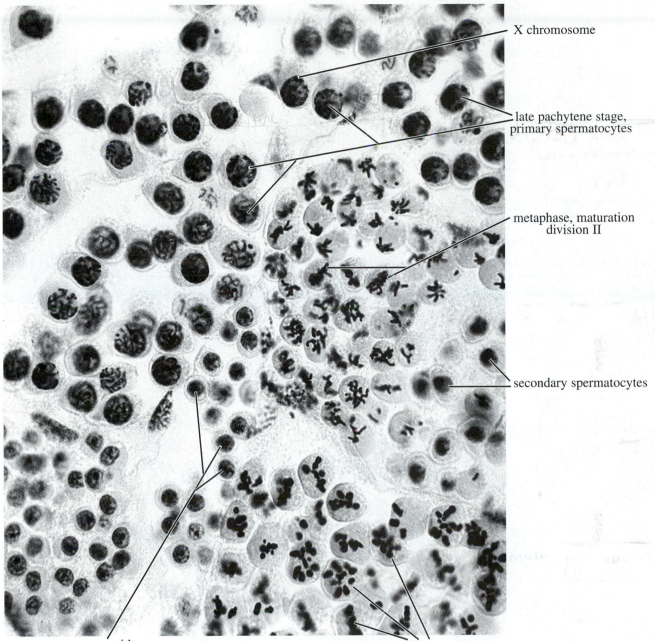

X chromosome

late pachytene stage, primary spermatocytes

metaphase, maturation division II

secondary spermatocytes

spermatids

metaphase, maturation division I

Figure 1.13

Grasshopper testis, longitudinal section showing area from Figure 1.10, Feulgen stain for DNA (725X).

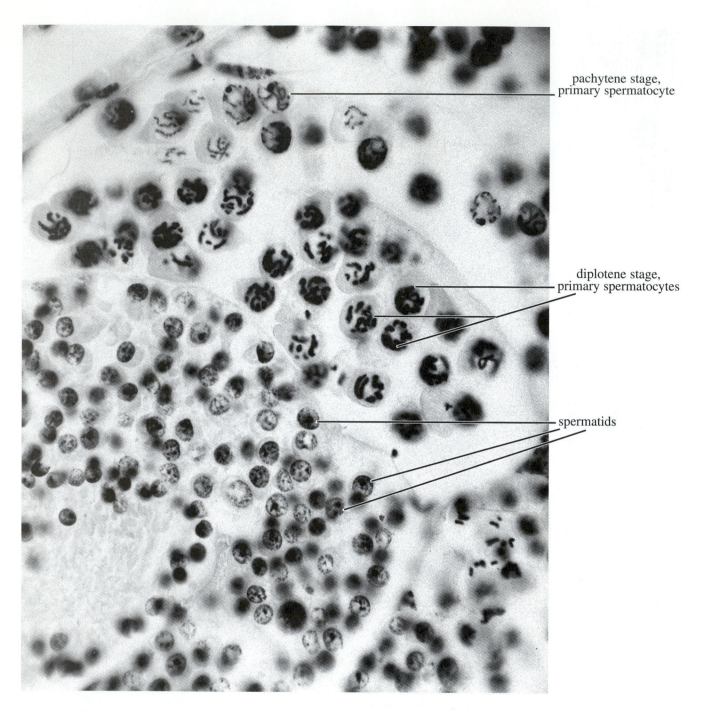

pachytene stage,
primary spermatocyte

diplotene stage,
primary spermatocytes

spermatids

Figure 1.14

Grasshopper testis, longitudinal section of testicular lobe, Feulgen stain for DNA (675X).

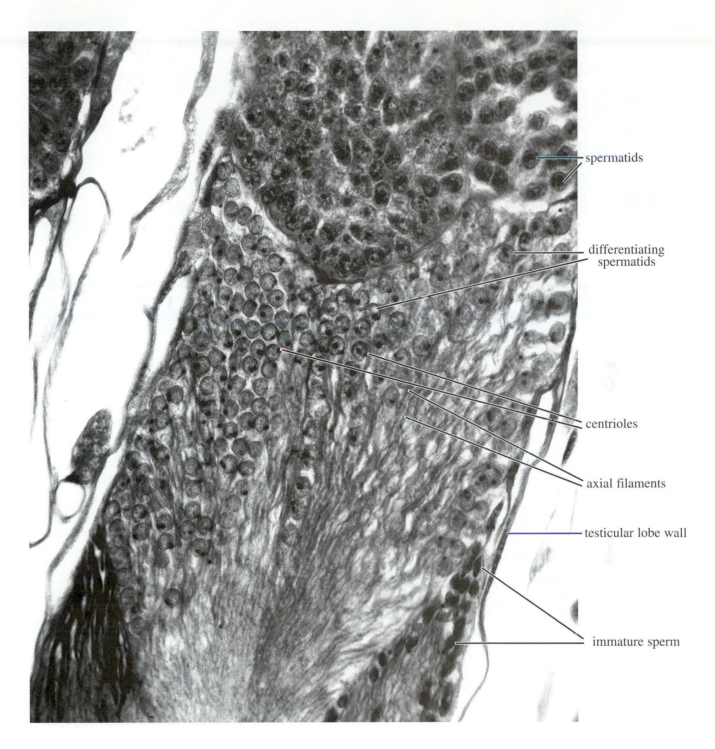

spermatids

differentiating
spermatids

centrioles

axial filaments

testicular lobe wall

immature sperm

Figure 1.15

Grasshopper testis, longitudinal section of testicular lobe, iron hematoxylin stain (700X).

Meiosis I

Chapter 2

Fertilization

Ascaris: **Figures 2.1-2.8**

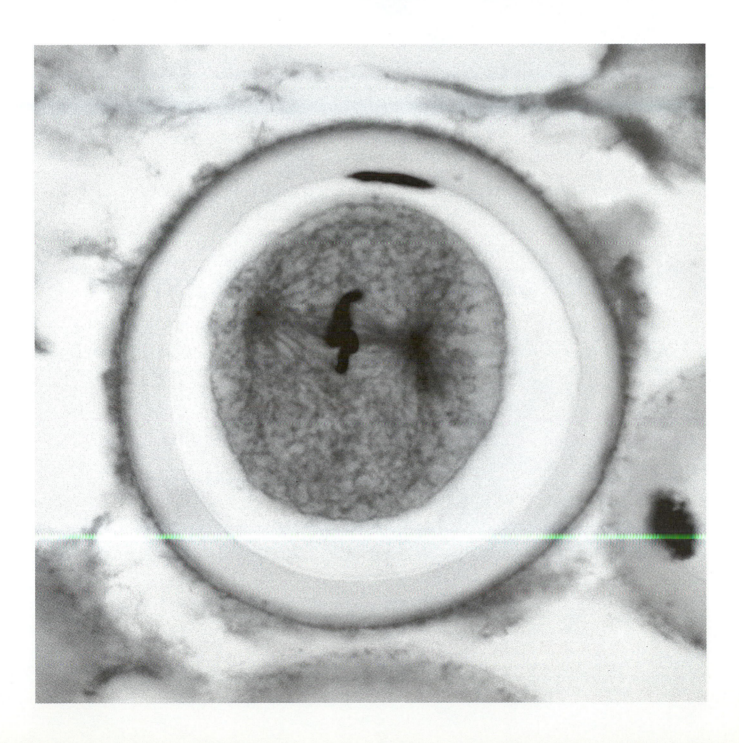

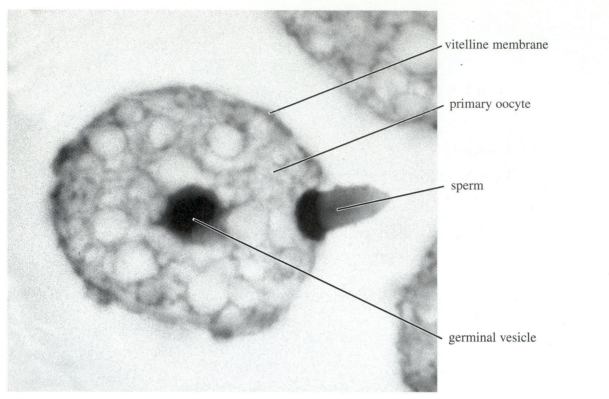

Figure 2.1

Ascaris sperm penetration stage (900X).

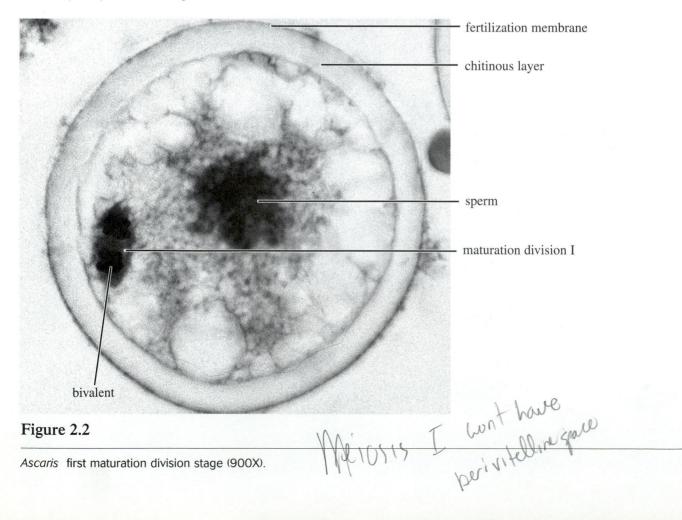

Figure 2.2

Ascaris first maturation division stage (900X).

Meiosis I wont have perivitelline space

Meosis II

⑤

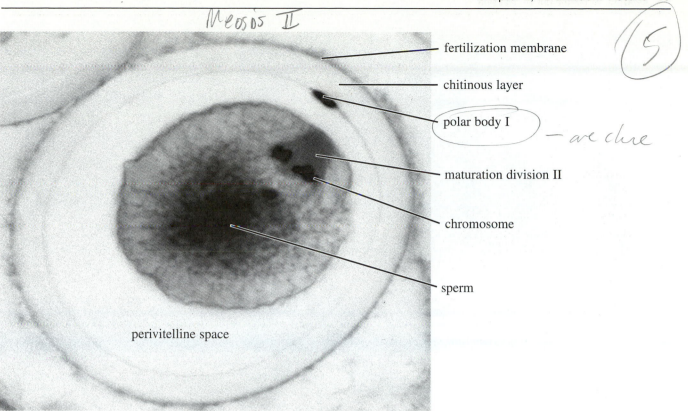

— are clue

fertilization membrane

chitinous layer

polar body I

maturation division II

chromosome

sperm

perivitelline space

Figure 2.3

Ascaris maturation division stage (900X).

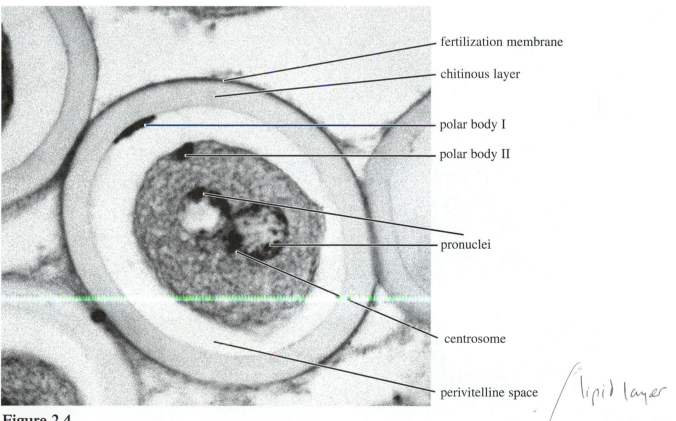

fertilization membrane

chitinous layer

polar body I

polar body II

pronuclei

centrosome

perivitelline space

lipid layer

Figure 2.4

Ascaris pronuclear stage (900X).

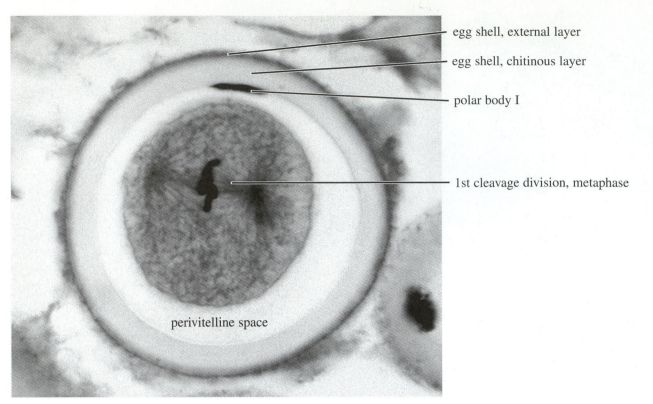

egg shell, external layer

egg shell, chitinous layer

polar body I

1st cleavage division, metaphase

perivitelline space

Figure 2.5

Ascaris first cleavage division stage (900X).

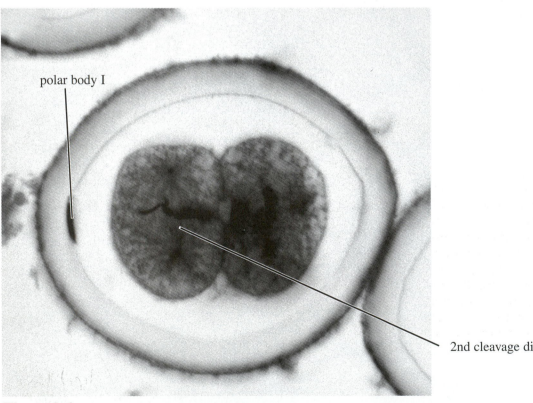

polar body I

2nd cleavage division, metaphase

Figure 2.6

Ascaris second cleavage division stage (900X).

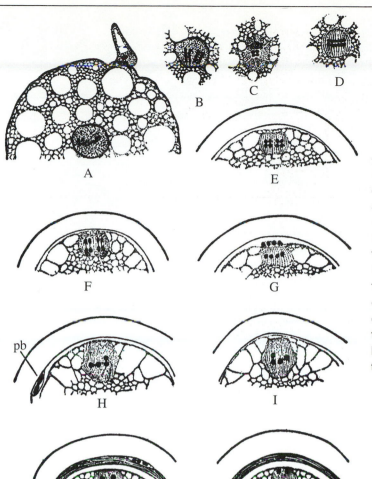

Figure 2.7

Formation of the polar bodies in *Ascaris megalocephala*, var. *bivalens* (Boveri). A, the egg with the sperm just entering; the germinal vesicle contains two rod-shaped tetrads (only one clearly shown), the number of chromosomes in earlier divisions having been four; B, the tetrads seen in profile; C, the tetrads seen in end view; D, first spindle forming (in this case inside the germinal vesicle); E, first polar spindle; F, tetrads dividing; G, first polar body formed containing, like the egg, two dyads; H, I, the dyads rotating into position for the second division; J, the dyads dividing; K, each dyad has divided into two single chromosomes, completing the reduction; pb, polar body I.

Figure 2.8

Fertilization of the egg of *Ascaris megalocephala*, var. *bivalens* (Boveri). A, the sperm has entered the egg (sperm nucleus is indicated by m); above are the closing phases in the formation of the second polar body (two chromosomes in each nucleus); B, the two pronuclei (m, male; f, female) in the reticular stage; the sphere (a) contains the dividing central body; C, chromosomes forming in the pronuclei; the central body divided; D, each pronucleus resolved into two chromosomes; sphere (a) has doubled; E, mitotic figure forming for the first cleavage; the chromosomes have already split; F, first cleavage in progress, showing the divergence of the daughter chromosomes toward the spindle poles (only three chromosomes are shown).

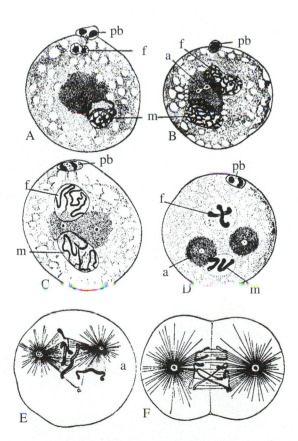

Chapter 3

Worm Development

C. elegans: **Figures 3.1-3.26**

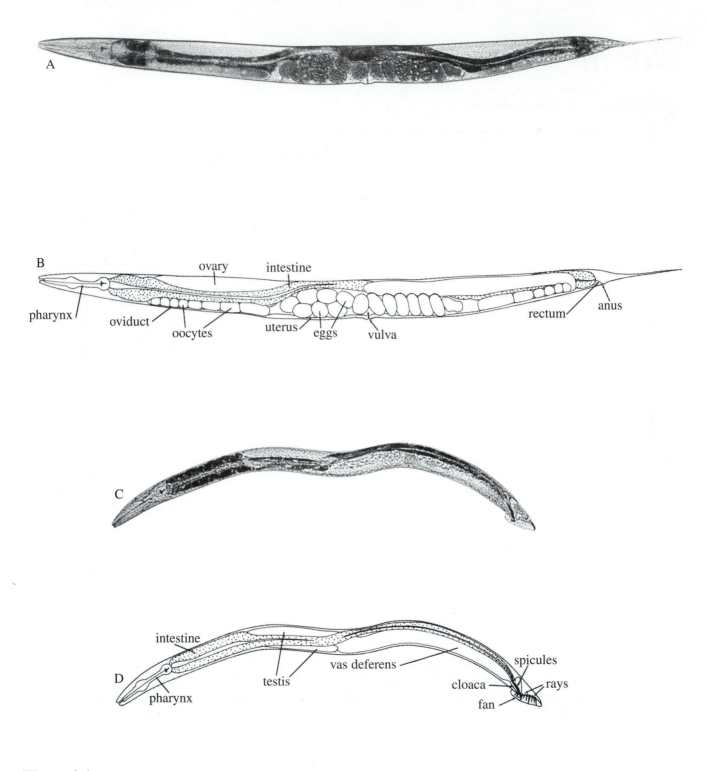

Figure 3.1

C. elegans adult worms, lateral views. A, photograph of a hermaphodite; B, diagram of the animal shown in A; C, photograph of a male; D, diagram of the animal shown in C. x150.

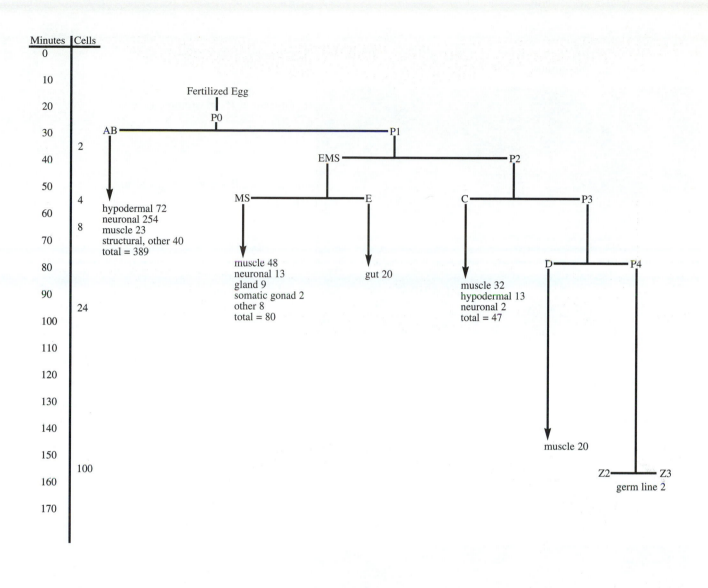

Figure 3.2

Diagram showing the lineage pattern of early cleavages in *C. elegans* embryos. The derivations of the six founder cells (AB, MS, E, C, D and P4) and the number of cells of each type present at hatching are indicated. Minutes of development at 25°C after fertilization are indicated at the left, as is the total number of cells generated by various time periods.

vitelline membrane

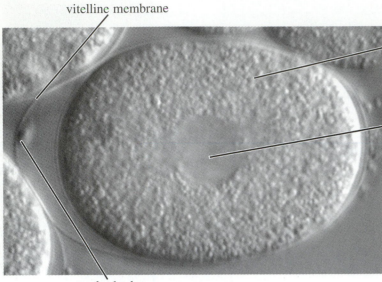

cytoplasm of P0 blastomere (zygote)

nucleus of P0 blastomere (zygote)

polar body

Figure 3.3

C. elegans embryo, zygote stage. 1800X.

nucleus of AB blastomere

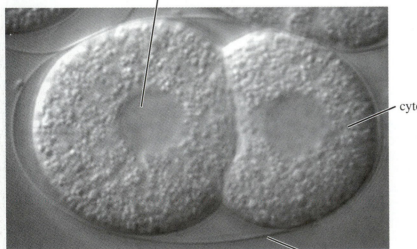

cytoplasm of P1 blastomere

vitelline membrane

Figure 3.4

C. elegans embryo, 2 cell stage. 1800X.

cleavage furrow   vitelline membrane

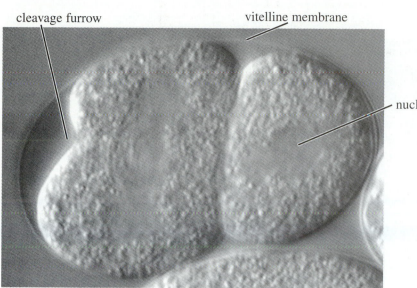

nucleus of P1 blastomere

Figure 3.5

C. elegans embryo, early 3-cell stage. 1800X.

cytoplasm of ABa (anterior) blastomere   cytoplasm of ABp (posterior) blastomere

cytoplasm of P1 blastomere

Figure 3.6

C. elegans embryo, 3-cell stage. 1800X.

cytoplasm of ABa blastomere   cytoplasm of ABp blastomere

cytoplasm of P2 blastomere

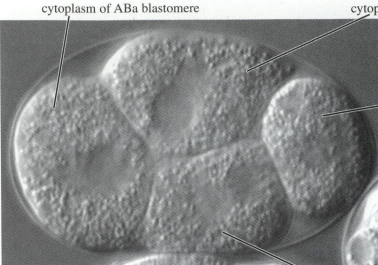

cytoplasm of EMS blastomere

Figure 3.7

C. elegans embryo, 4-cell stage. 1800X.

cytoplasm of ABa left and right blastomeres

nucleus of ABp blastomere

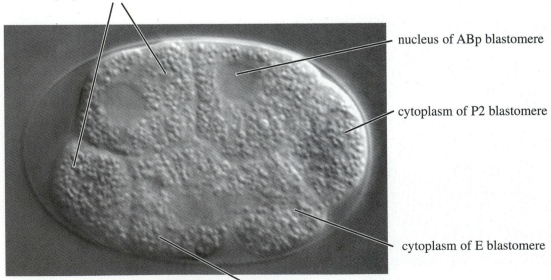

cytoplasm of P2 blastomere

cytoplasm of E blastomere

cytoplasm of MS blastomere

Figure 3.8

C. elegans embryo, 6-cell stage. 1800X.

cytoplasm of ABa left and right blastomeres

cytoplasm of ABp left and right blastomeres

cytoplasm of C blastomere

cytoplasm of P3 blastomere

cytoplasm of E blastomere

cytoplasm of MS blastomere

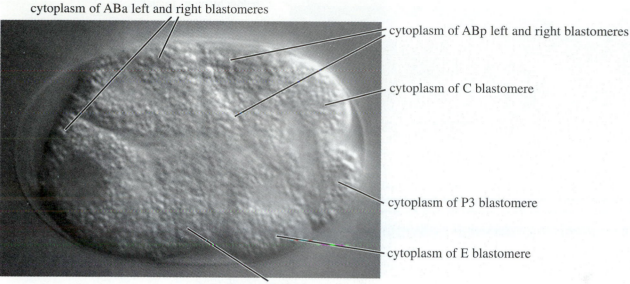

Figure 3.9

C. elegans embryo, 8-cell stage. 1800X.

AB blastomeres (8 present but only 5 shown)

position of C blastomere (located below plane of focus)

cytoplasm of P3 blastomere

nucleus of E blastomere

nucleus of MS blastomere

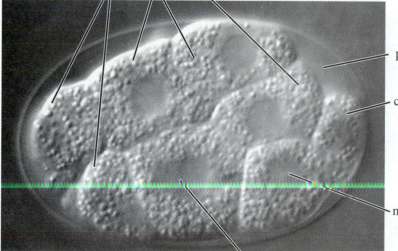

Figure 3.10

C. elegans embryo, 12-cell stage. 1800X.

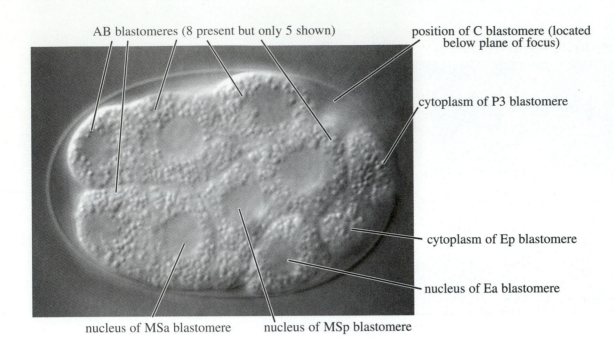

AB blastomeres (8 present but only 5 shown)

position of C blastomere (located below plane of focus)

cytoplasm of P3 blastomere

cytoplasm of Ep blastomere

nucleus of Ea blastomere

nucleus of MSa blastomere

nucleus of MSp blastomere

Figure 3.11

C. elegans embryo, 14-cell stage. 1800X.

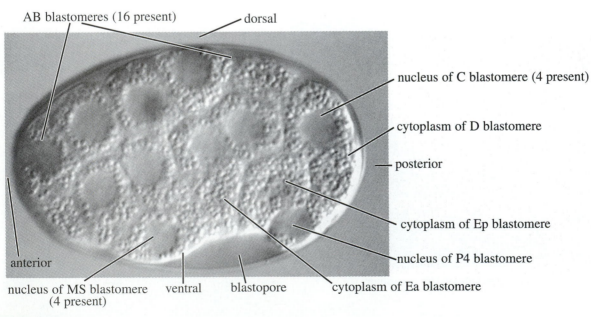

AB blastomeres (16 present)

dorsal

nucleus of C blastomere (4 present)

cytoplasm of D blastomere

posterior

cytoplasm of Ep blastomere

nucleus of P4 blastomere

anterior

nucleus of MS blastomere (4 present)

ventral

blastopore

cytoplasm of Ea blastomere

Figure 3.12

C. elegans embryo, 28-cell stage. Beginning of gastrulation. about 2 hours after fertilization. 1800X.

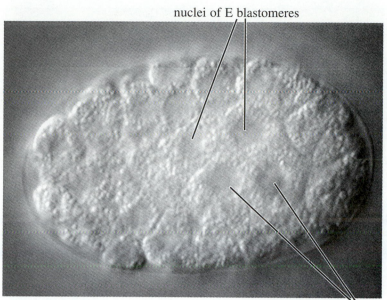

nuclei of E blastomeres

nuclei of E blastomeres

Figure 3.13

C. elegans embryo, 4E stage (44-50 cells). 1800X.

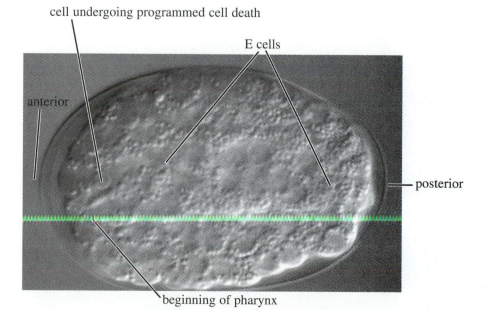

cell undergoing programmed cell death

E cells

anterior

posterior

beginning of pharynx

Figure 3.14

C. elegans embryo, 8E stage. 1800X.

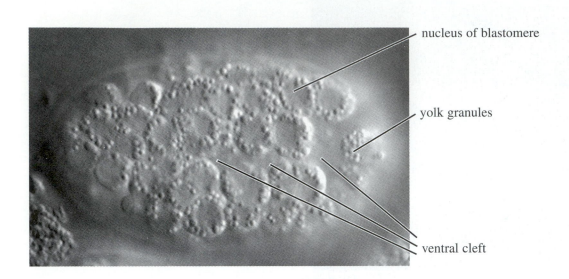

nucleus of blastomere

yolk granules

ventral cleft

Figure 3.15

C. elegans embryo, just prior to ventral cleft closure. Ventral view. 1800X.

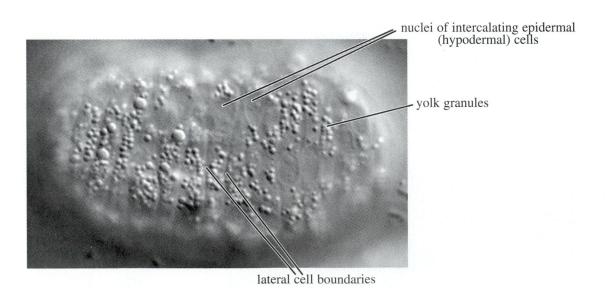

nuclei of intercalating epidermal
(hypodermal) cells

yolk granules

lateral cell boundaries

Figure 3.16

C. elegans embryo, during dorsal epidermal (hypodermal) intercalation. Dorsal view, about 4 hours after fertilization.
1800X.

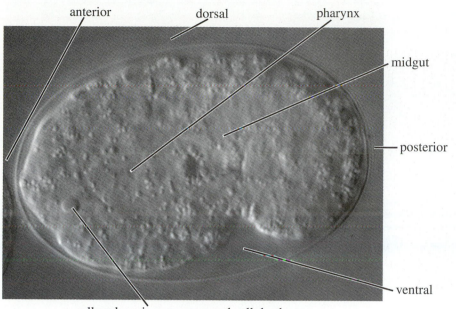

Figure 3.17

C. elegans embryo, ventral indentation stage. Lateral view, about 6 hours after fertilization. 1800X.

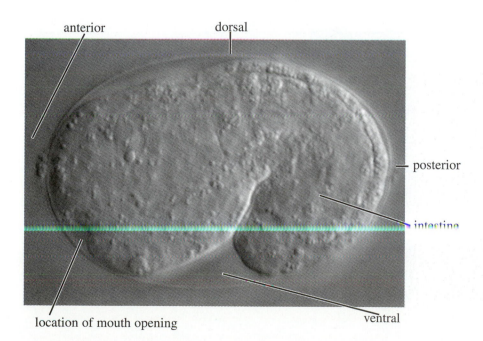

Figure 3.18

C. elegans embryo, comma stage. Lateral view 1800X.

cell undergoing programmed cell death   dorsal

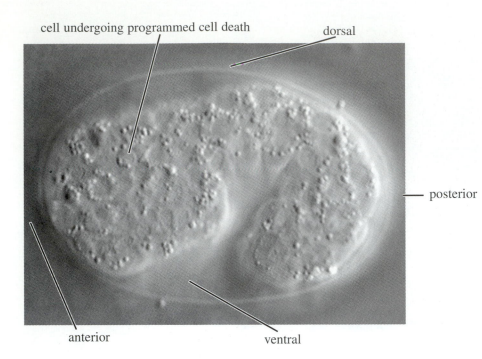

anterior   ventral

posterior

Figure 3.19

C. elegans embryo, comma stage, lateral view (same embryo as in Figure 3.18, but at a more superficial focal plane). 1800X.

anterior (head end)   vitelline membrane

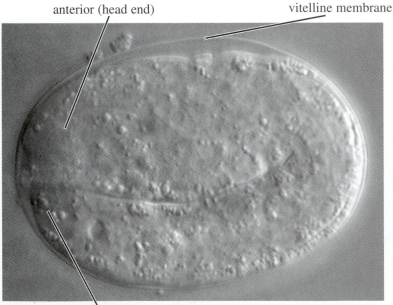

posterior (tail end)

Figure 3.20

C. elegans embryo, 2-fold stage. Lateral view. 1800X.

posterior (tail end)   anterior (head end)

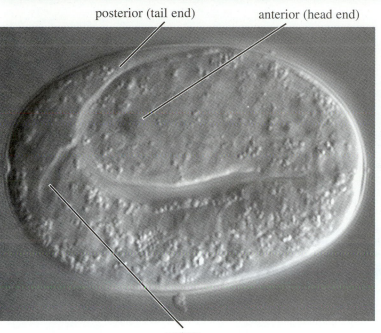

rectum

Figure 3.21

C. elegans embryo, early 3-fold stage. Lateral view. 1800X.

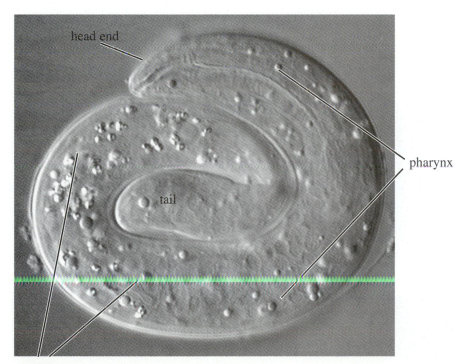

head end

pharynx

tail

midgut

Figure 3.22

C. elegans embryo, 3-fold stage. Lateral view. 1800X.

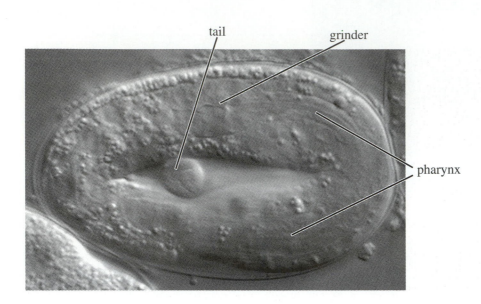

Figure 3.23

C. elegans embryo, about to hatch. About 10 hours after fertilization. 1800X.

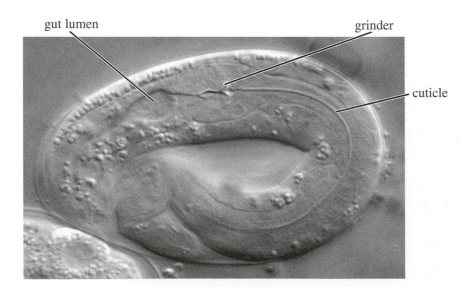

Figure 3.24

C. elegans embryo, about to hatch. About 10 hours after fertilization. 1800X.

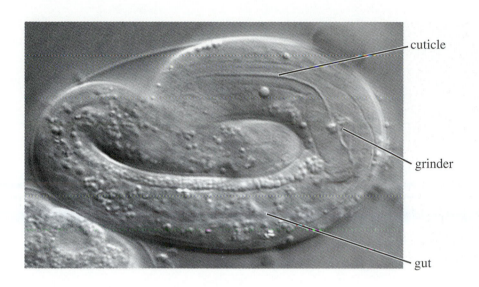

cuticle

grinder

gut

Figure 3.25

C. elegans embryo, about to hatch. About 10 hours after fertilization. 1800X.

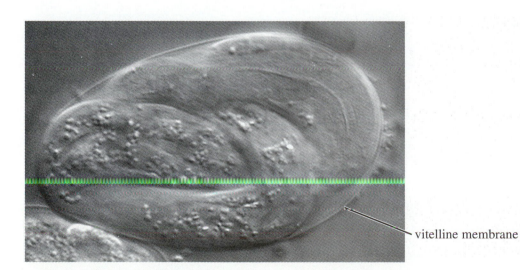

vitelline membrane

Figure 3.26

C. elegans embryo, about to hatch. About 10 hours after fertilization. 1800X.

Chapter 4

Echinoderm Development

Sea Urchin: **Figures 4.1-4.19**
Starfish: **Figures 4.20-4.32**

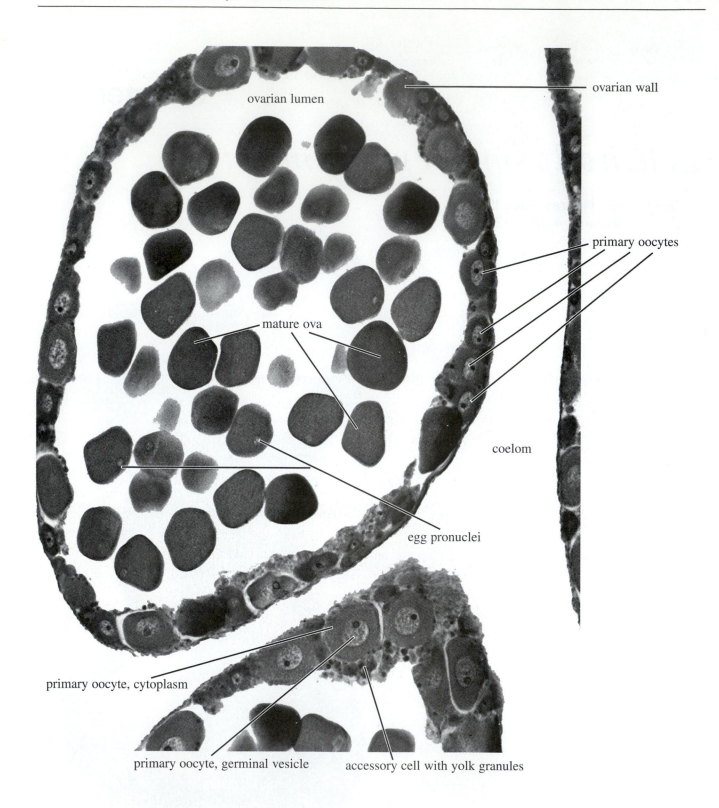

Figure 4.1

Sea urchin ovary, *Arbacia* (290X).

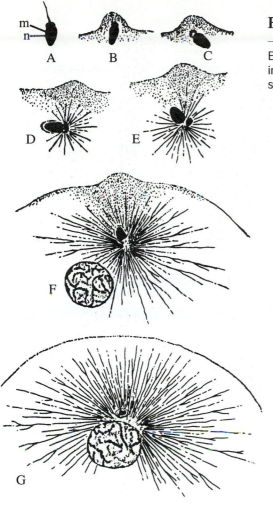

Figure 4.2

Entrance and rotation of the sperm head and formation of the sperm aster in the sea urchin *Toxopneustes.* m, mitochondria of sperm; n, nucleus of sperm (A-E, 1600X; F, G, 800X).

Figure 4.3

Conjugation of the gamete nuclei and division of the sperm aster in the sea urchin *Toxopneustes.* The smaller circular body is the sperm pronucleus resting upon the larger egg pronucleus (1000X).

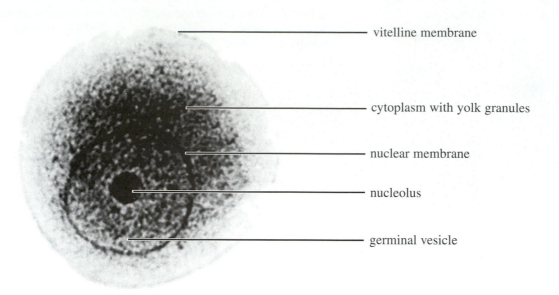

vitelline membrane

cytoplasm with yolk granules

nuclear membrane

nucleolus

germinal vesicle

Figure 4.4

Sea urchin primary oocyte (550X).

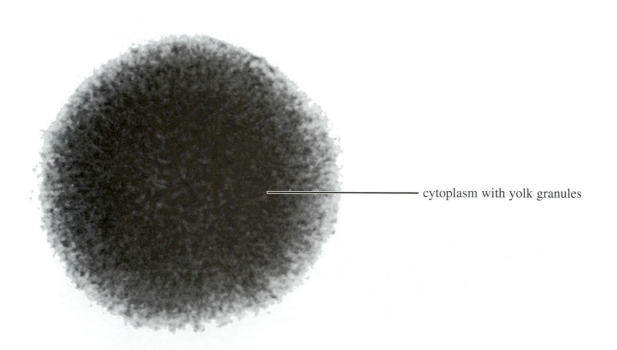

cytoplasm with yolk granules

Figure 4.5

Sea urchin fertilized egg (550X).

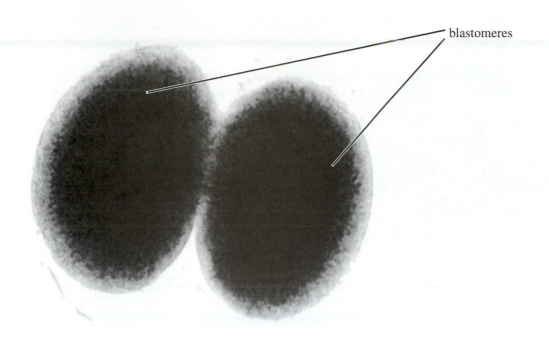

blastomeres

Figure 4.6

Sea urchin two cells (550X).

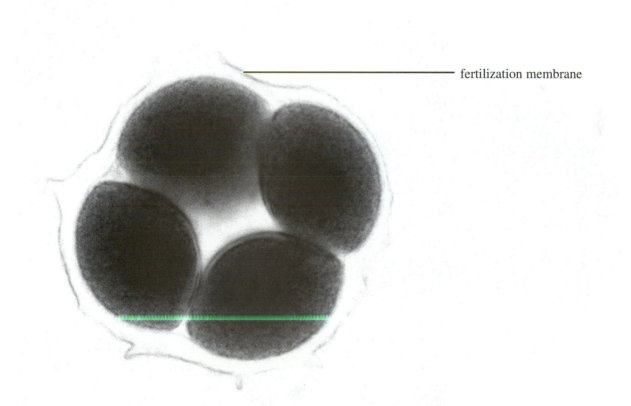

fertilization membrane

Figure 4.7

Sea urchin four cells (550X).

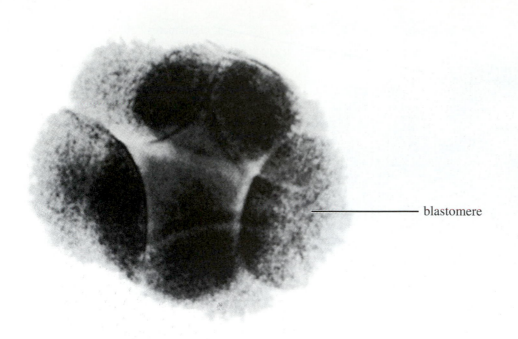

blastomere

Figure 4.8

Sea urchin eight cells (550X).

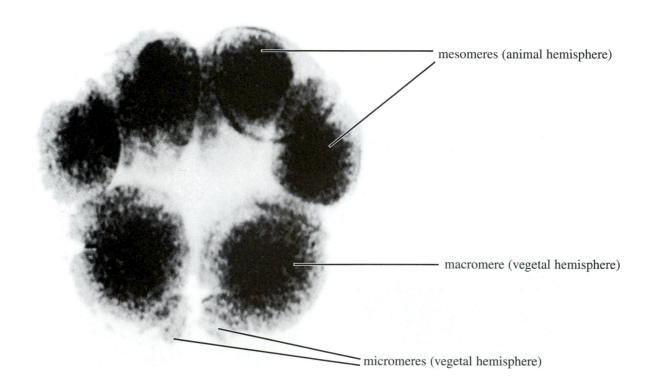

mesomeres (animal hemisphere)

macromere (vegetal hemisphere)

micromeres (vegetal hemisphere)

Figure 4.9

Sea urchin sixteen cells (550X).

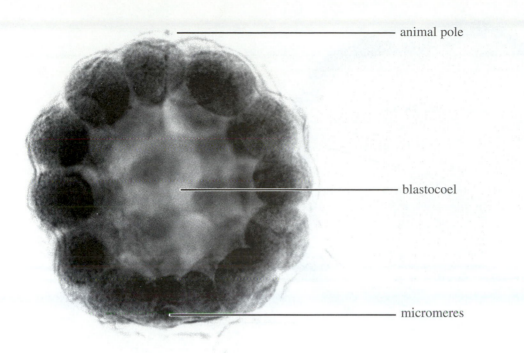

animal pole

blastocoel

micromeres

Figure 4.10

Sea urchin early blastula (550X).

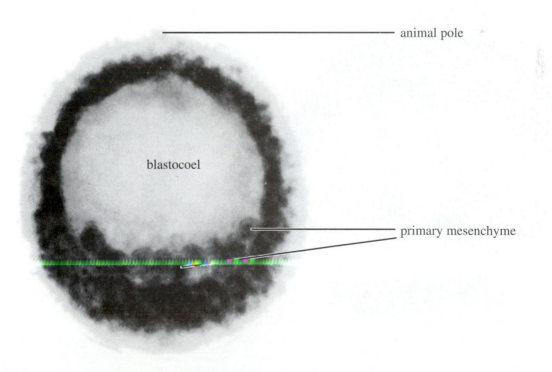

animal pole

blastocoel

primary mesenchyme

Figure 4.11

Sea urchin late blastula (550X).

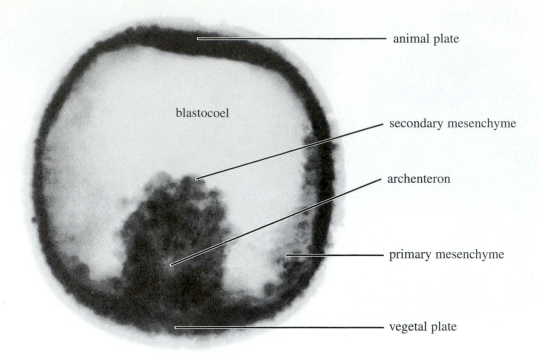

animal plate

blastocoel

secondary mesenchyme

archenteron

primary mesenchyme

vegetal plate

Figure 4.12

Sea urchin early gastrula (550X).

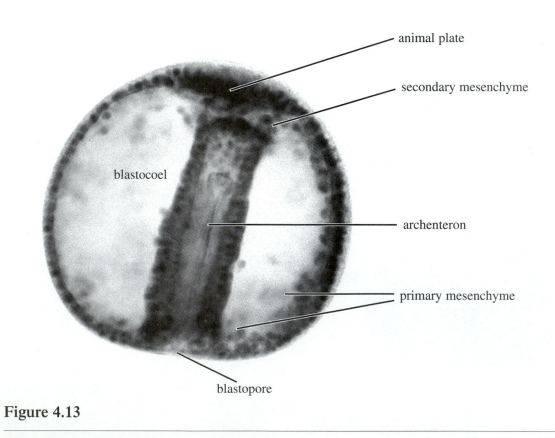

animal plate

secondary mesenchyme

blastocoel

archenteron

primary mesenchyme

blastopore

Figure 4.13

Sea urchin late gastrula (550X).

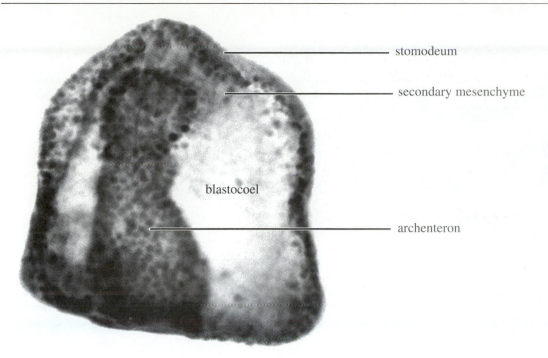

stomodeum

secondary mesenchyme

blastocoel

archenteron

Figure 4.14

Sea urchin prism larva (550X).

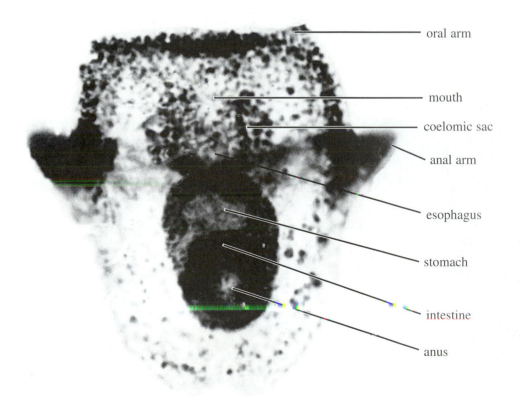

oral arm

mouth

coelomic sac

anal arm

esophagus

stomach

intestine

anus

Figure 4.15

Sea urchin early pluteus larva (550X).

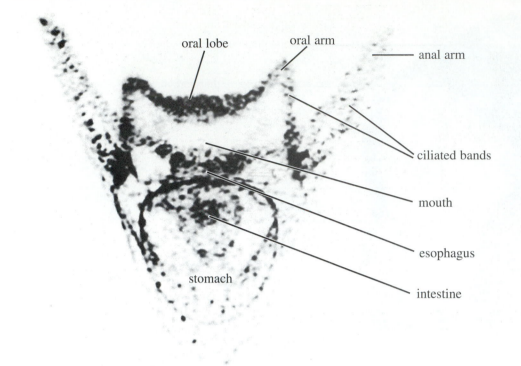

Figure 4.16

Sea urchin late pluteus larva, ventral view (550X).

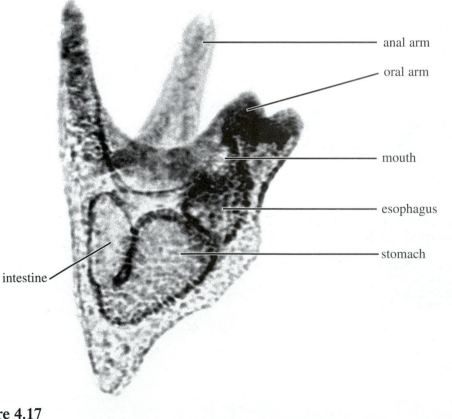

Figure 4.17

Sea urchin late pluteus larva, lateral view (550X).

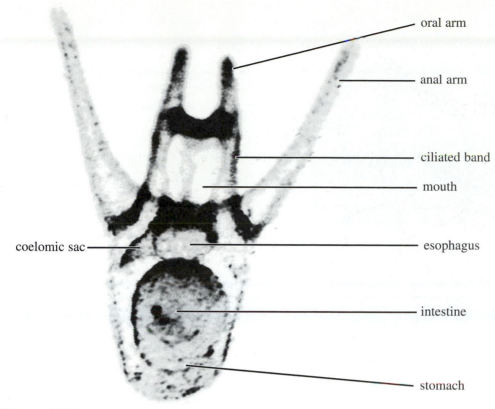

oral arm

anal arm

ciliated band

mouth

coelomic sac

esophagus

intestine

stomach

Figure 4.18

Mature pluteus larva, sea urchin *Arbacia* (550X).

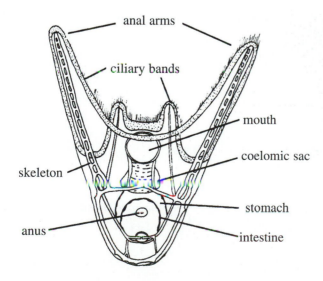

anal arms

ciliary bands

mouth

coelomic sac

skeleton

stomach

anus

intestine

Figure 4.19

Mature pluteus larva, sea urchin *Tripneustes gratilla* (550X).

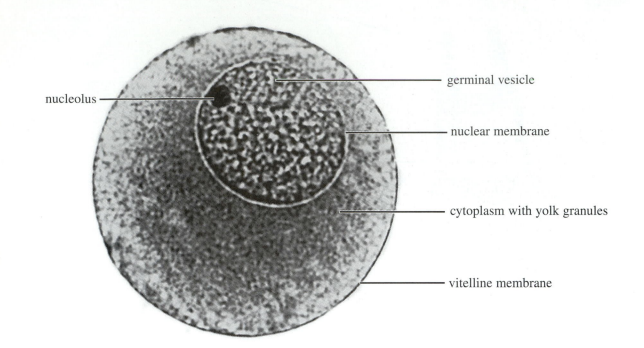

nucleolus

germinal vesicle

nuclear membrane

cytoplasm with yolk granules

vitelline membrane

Figure 4.20

Star fish primary oocyte (550X).

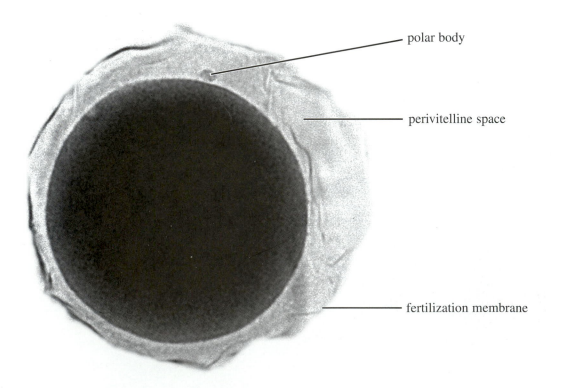

polar body

perivitelline space

fertilization membrane

Figure 4.21

Star fish fertilized egg (550X).

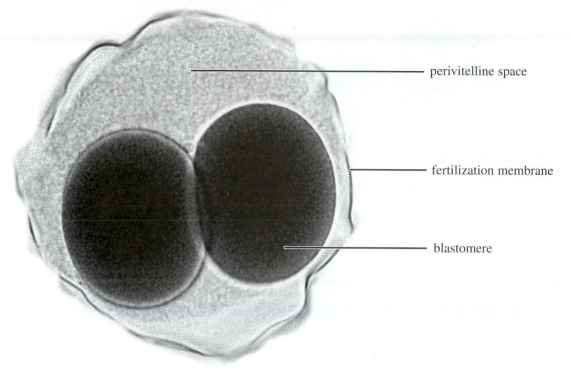

perivitelline space

fertilization membrane

blastomere

Figure 4.22

Star fish two cells (550X).

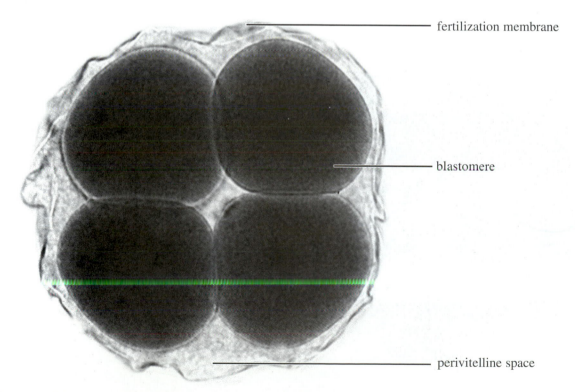

fertilization membrane

blastomere

perivitelline space

Figure 4.23

Star fish four cells (550X).

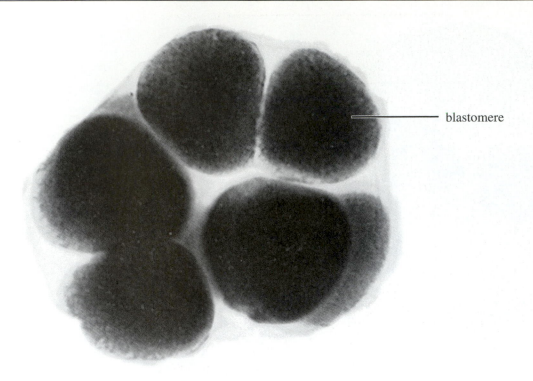

blastomere

Figure 4.24

Star fish eight cells (550X).

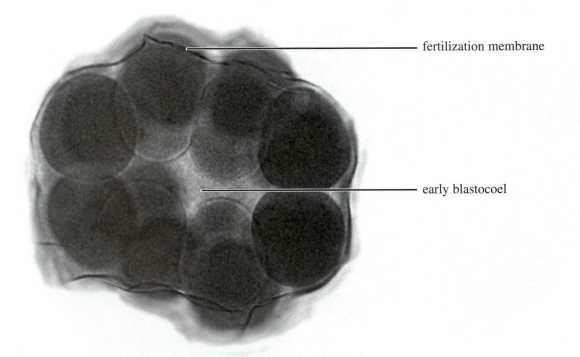

fertilization membrane

early blastocoel

Figure 4.25

Star fish sixteen cells (550X).

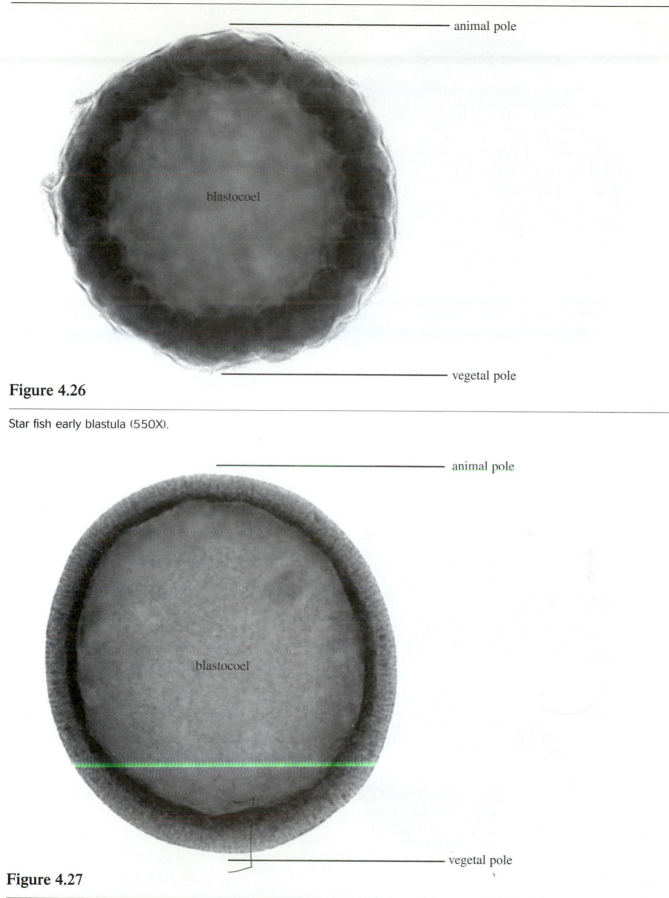

animal pole

blastocoel

vegetal pole

Figure 4.26

Star fish early blastula (550X).

animal pole

blastocoel

vegetal pole

Figure 4.27

Star fish late blastula (550X).

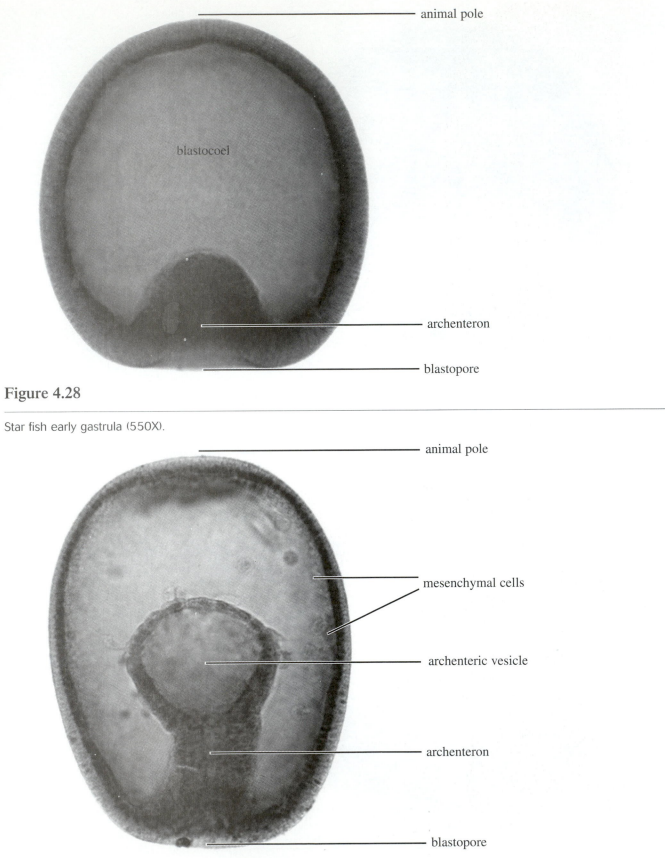

animal pole

blastocoel

archenteron

blastopore

Figure 4.28

Star fish early gastrula (550X).

animal pole

mesenchymal cells

archenteric vesicle

archenteron

blastopore

Figure 4.29

Star fish mid-gastula (550X).

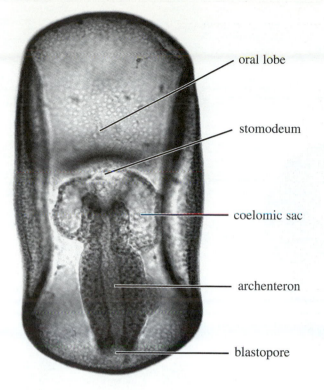

oral lobe

stomodeum

coelomic sac

begin to be more defined

archenteron

blastopore

Figure 4.30

Star fish late gastula, ventral view (550X).

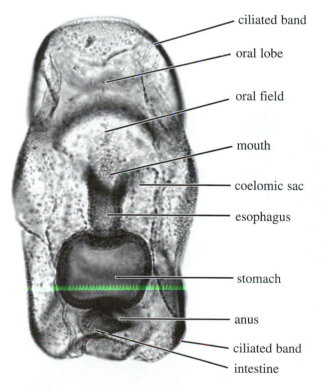

ciliated band

oral lobe

oral field

mouth

coelomic sac

esophagus

stomach

anus

ciliated band

intestine

Figure 4.31

Star fish bipinnaria larva, ventral view (550X).

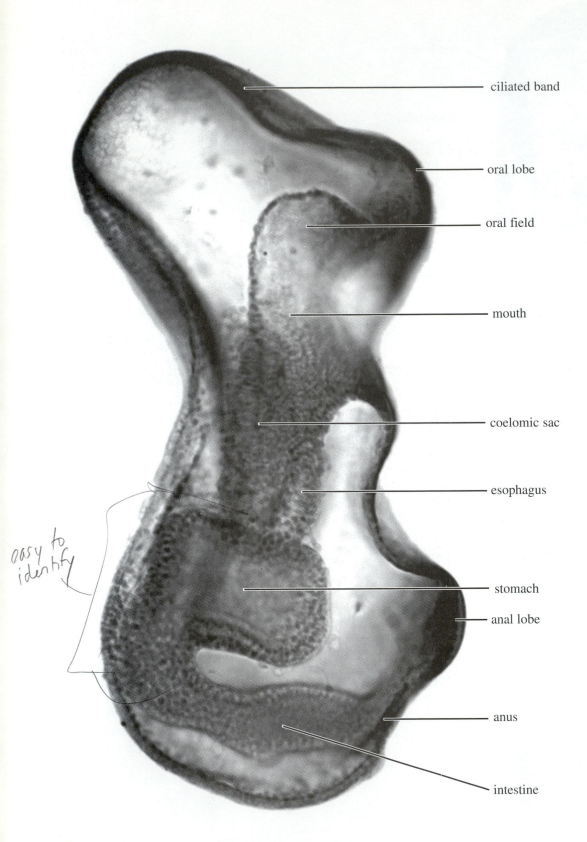

ciliated band

oral lobe

oral field

mouth

coelomic sac

esophagus

stomach

anal lobe

anus

intestine

easy to identify

Figure 4.32

Star fish bipinnaria larva, lateral view (1100X).

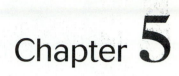

Chordate Development

Amphioxus: **Figures 5.1-5.20**

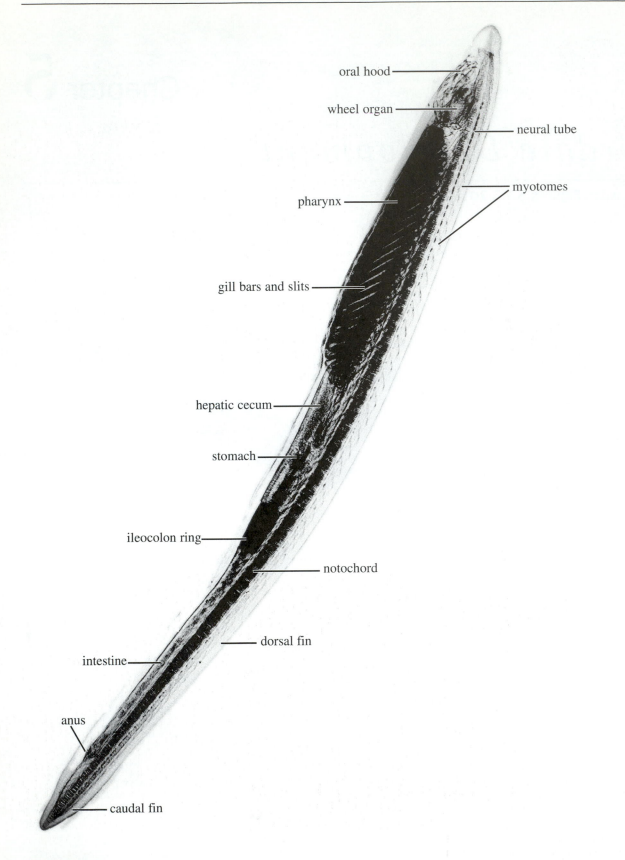

oral hood

wheel organ

neural tube

myotomes

pharynx

gill bars and slits

hepatic cecum

stomach

ileocolon ring

notochord

intestine

dorsal fin

anus

caudal fin

Figure 5.1

Amphioxus, immature adult (40X).

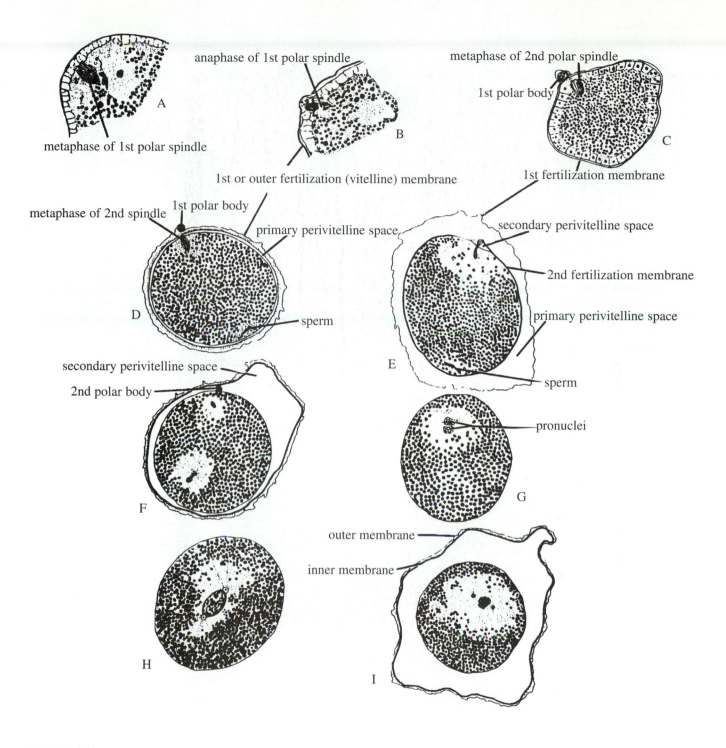

Figure 5.2

Fertilization of the egg of *Amphioxus*. A, metaphase of the first maturation division before sperm entrance; B, anaphase of the first maturation division before sperm entrance; C, first polar body and metaphase of the second maturation division before sperm entrance; D, sperm has pentrated egg near its vegetal pole; E, outer vitelline membrane has separated from egg; the second polar body is forming; F, the inner vitelline membrane has lifted and is fused with the first vitelline membrane to form the 2nd fertilization membrane; the pronuclei of the egg and sperm have formed; the sperm aster and second polar body are present; G, the sperm and egg pronucli have met; H, pronuclei fused; I, prophase of the first cleavage mitosis.

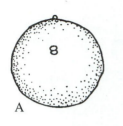

A

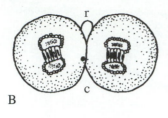

B

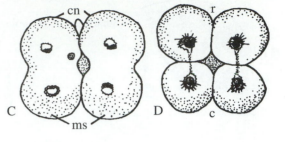

C

D

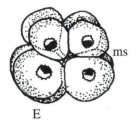

E

F

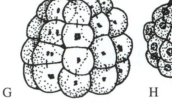

G

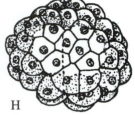

H

I

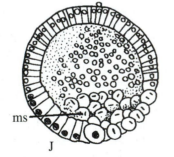

J

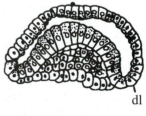

K

L

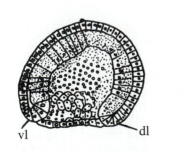

M

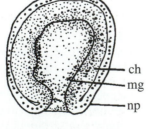

N

O

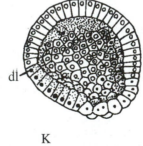

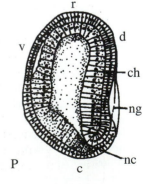

P

Figure 5.3

(opposite) *Amphioxus* development, fertilized egg to late embryo stages. A, egg one hour after fertilization; the two pronuclei are just above the center, and the second polar body marks the animal pole. B, the 2-cell stage with the second cleavage in progress 2 hours after fertilization; C, second cleavage in telophase 2 hours after fertilization; the mesodermal crescent is in the caudal region and the chorda-neural crescent is opposite; D, vegetal view of the 4-cell stage 2 hours after fertilization; the rostral cells are slightly larger than the caudal cells; E, 2.5 hours after fertilization; the third cleavage has divided the cells unequally yielding 4 smaller micromeres and 4 larger macromeres; the caudal macromeres contain the mesoderm crescent; F, the 16-cell stage 2.5 hours after fertilization; a definite blastocoel has formed; G, the 32-cell stage 2.5 hours after fertiliztion; the blastocoel is open at the vegetal pole; H, the 64-cell stage 3.25 hours after fertilization; the vegetal cells are larger than the animal cells; the blastocoel is now closed; I, the 256-cell stage 4 hours after ferilization; the lower large cells will form the endoderm with the dorsal lip at their border; J, the blastula 5.5 hours after fertilization seen in optical section; the small dividing cells at the caudal surface were derived from the mesodermal crescent; the larger caudal cells are prospective endoderm; the second polar body marks the animal pole; K, a blastula 5.5 hours after fertilization in optical section; the endodermal cells have flattened with the future dorsal lip at one margin and the mesodermal crescent cells at the other; L, early gastrula about 8 hours after fertilization in optical section; the endoderm has invaginated into the blastocoel; the dorsal lip of the blastopore is on the right and contains cells of the future notochord; M, gastrula of 11 hours in optical section; the blastocoel has been nearly filled by the invaginated archenteron; the dorsal lip of the blastopore is on the right; mesodermal cells are still involuting over the ventral lip; N, gastrula of 13 hours in optical section; the embryo is elongating and the blastopore is constricting; the dorsal wall of the embryo on the right consists of the ectodermal neural plate underlain by notochord and mesodermal groove, both now in the wall of the archenteron; O, embryo of 15-16 hours seen from the dorsal side; the flattened area is the neural plate, depressed caudally as the neural groove; a remnant of the blastocoel persists; the margin of the archenteron (gastrocoel) is indicated by a dotted line; the blastopore is nearly closed and the neural groove is being covered by ectodermal folds derived from the ventral and lateral lips of the blastopore; P, embryo of 15 hours showing the ectodermal overgrowth of the neural goove; a neurenteric canal connects the neural groove with the blastopore; the notochord underlies the neural groove.

bc, blastocoel
c, caudal
ch, notochord
cn, chorda-neural crescent
d, dorsal
dl, dorsal lip
gc, gastrocoel (archenteron)
ms, mesodermal crescent

nc, neurenteric canal
ng, neural groove
np, neural plate
l, left
r, rostral
rt, right
v, ventral
vl, ventral lip

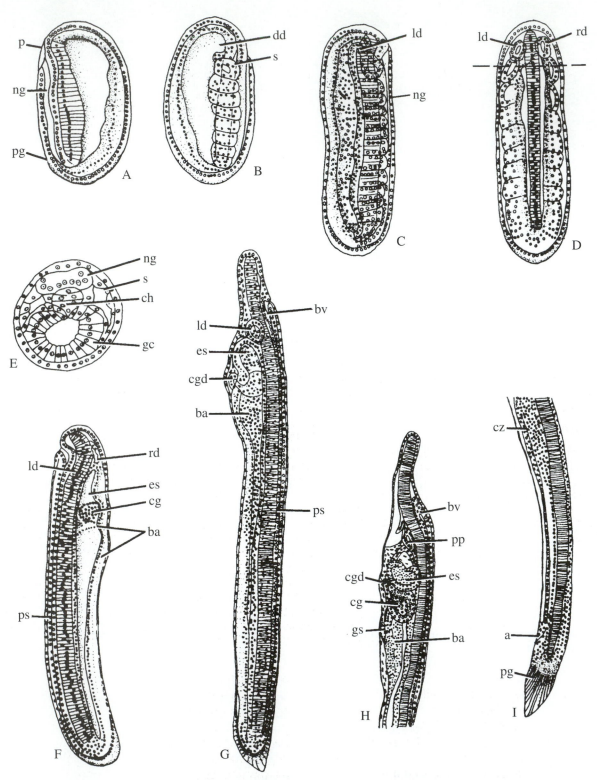

Figure 5.4

(opposite) *Amphioxus* development, late embryo to larval stages. A, embryo of 16 hours viewed from the right side; the ecto-dermal overgrowth of the neural groove is complete except for the open neuropore at the rostral end of the neural groove; pigment granules are present at the caudal surface; B, embryo of 18 hours viewed from the left side; a series of paired somites has evaginated from the archenteron as has a dorsal diverticulum at its rostral end; C, embryo of 24 hours viewed from the left side; elongation is marked with 10 or 11 somites; the caudal half of the neural groove has closed forming a neural tube; a left dorsal diverticulum has evaginated from the archenteron; the embryo is now ciliated and hatches at this stage; D, embryo of 24 hours viewed from the dorsal side; the central rod is the notochord, bordered on each side by a row of somites; the most rostral vesicles are the right and left diverticula of the archenteron; the transverse line indi-cates the plane of section through the first somites shown in E; E, transverse section of a 24 hour embryo through the first somites as indicated in D; the notochord and somites have separated from the archenteron (gastrocoel); the neural groove overlies the notochord and somites; F, larva of 26 hours viewed from the right side; the right diverticulum is expanding to form the head cavity; the left diverticulum is visible through the notochord; the rudiment of the endostyle lies on the floor of the rostral gut; rostral to the branchial segment (pharynx) of the gut is the clubshaped gland; a light-sensitive pigment spot has formed in the neural tube; G, larva of 48 hours viewed from the left; the notochord now extends well beyond the brain vesicle; the left diverticulum, endostyle, branchial segment (pharynx) and pigment spot are as in F; the duct of the clubshaped gland is visible; H, rostral third of the larva of 96 hours viewed from the side; above the noto-chord lies the brain vesicle and below is the preoral pit; just caudal is the first segment of the gut into which the mouth opens; a dark heart-shaped endostyle and the clubshaped gland follows; below the branchial segment (pharynx) is the first gill slit; I, caudal third of the larva of 96 hours; a ciliated zone (ileocolon) ring of the intestine is forming; the intestine terminates at the anus on the ventral side, just rostral to the caudal fin.

a, anus
ba, branchial rudiment
bv, brain vesicle
cg, clubshaped gland
cgd, duct of clubshaped gland
ch, notochord
cz, ciliated zone of gut
dd, dorsal diverticulum
es, endostyle
gc, gastrocoel (archenteron)

gs, gill slit
ld, left diverticulum
ng, neural groove
p, neuropore
pg, pigment granules
pp, preoral pit
ps, pigment spot
rd, right dorsal diverticulum
s, somite

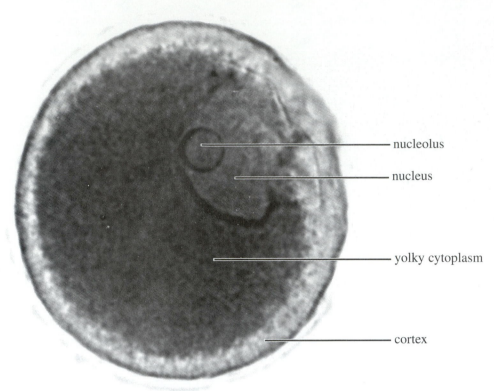

nucleolus

nucleus

yolky cytoplasm

cortex

Figure 5.5

Amphioxus oocyte (450X).

yolky cytoplasm

Figure 5.6

Amphioxus uncleaved egg (450X).

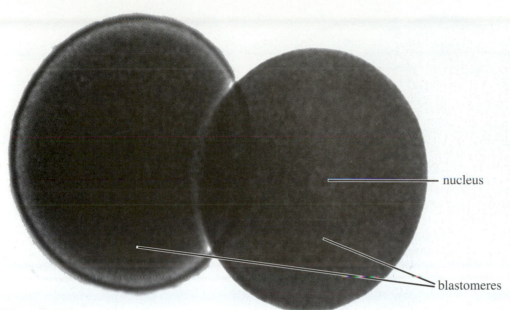

nucleus

blastomeres

Figure 5.7

Amphioxus two cells (450X).

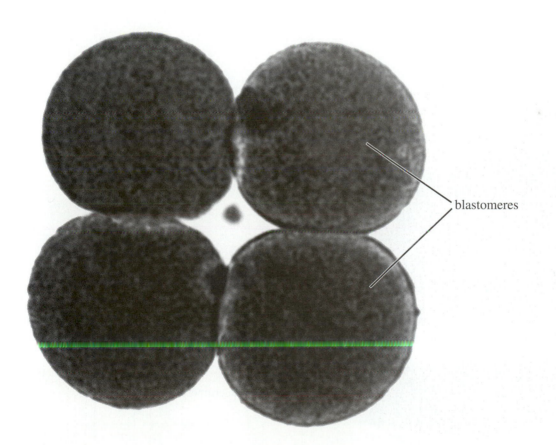

blastomeres

Figure 5.8

Amphioxus embryo, four cells; cells are in mitosis, so nuclei are not visible (450X).

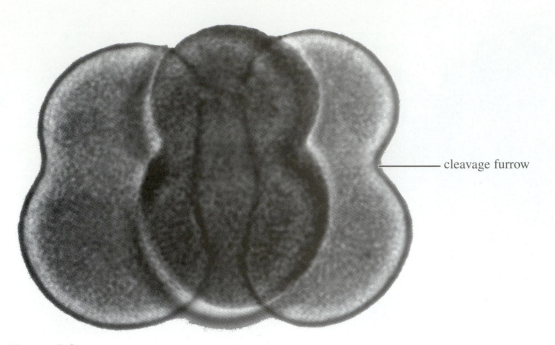

cleavage furrow

Figure 5.9

Amphioxus embryo, third cleavage; this cleavage division is not yet complete (450X).

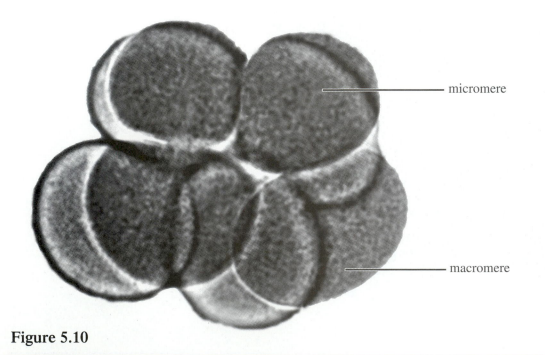

micromere

macromere

Figure 5.10

Amphioxus embryo, eight cells; cells are in mitosis, so nuclei are not visible (450X).

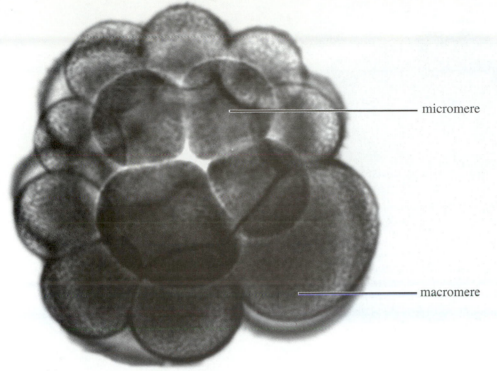

micromere

macromere

Figure 5.11

Amphioxus embryo, thirty two cells (450X).

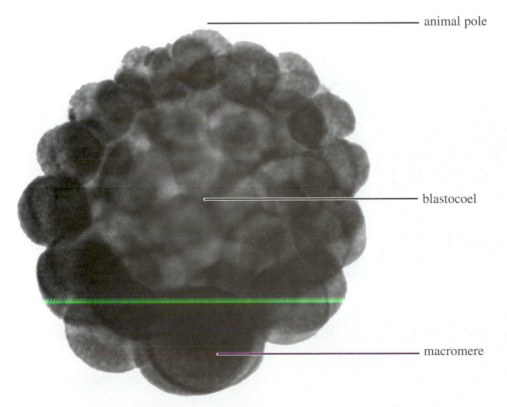

animal pole

blastocoel

macromere

Figure 5.12

Amphioxus embryo, early blastula viewed in optical section (450X).

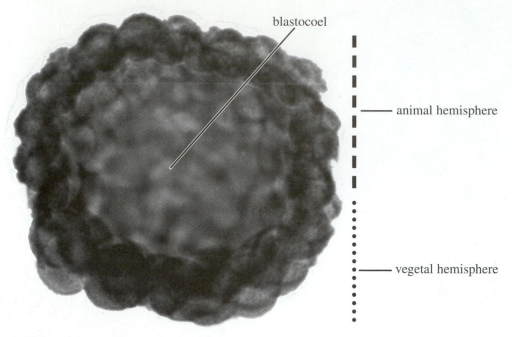

Figure 5.13

Amphioxus embryo, late blastula viewed in optical section; the larger vegetal cells will invaginate to form endoderm (450X).

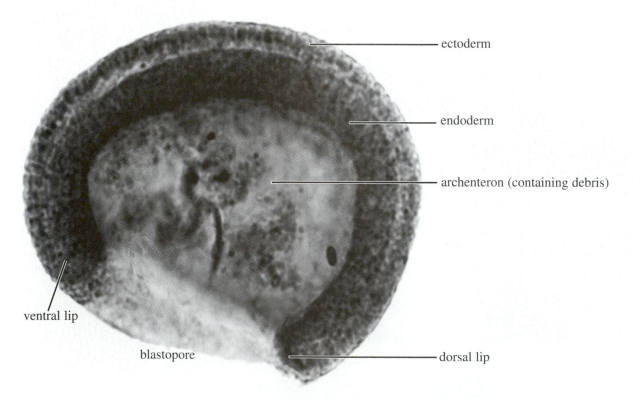

Figure 5.14

Amphioxus embryo, gastrula viewed in optical section; the invaginated endoderm and mesoderm have filled the blastocoel (450X).

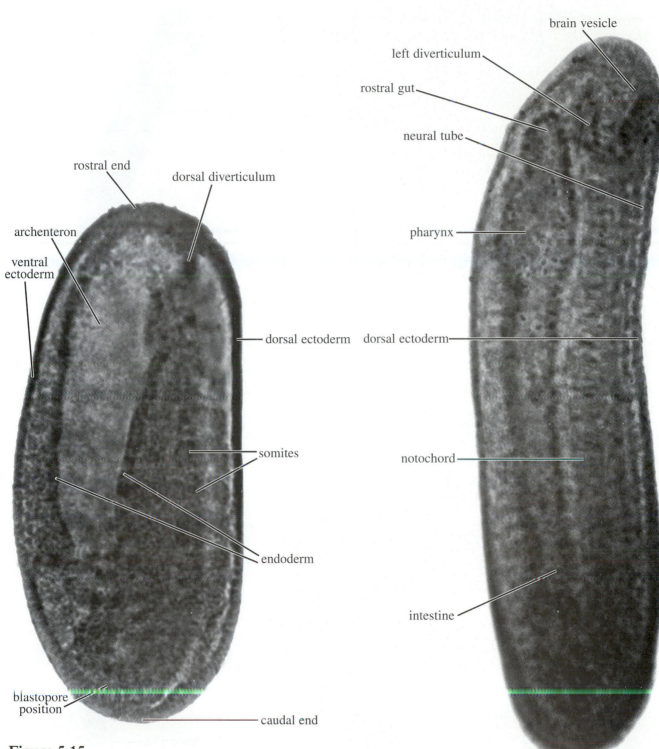

Figure 5.15

Amphioxus late embryo viewed in optical section; gastrulation is completed and the blastopore is nearly closed; embryo is elongating; notochord and somites are separating from the archenteron (450X).

Figure 5.16

Amphioxus early larva about 26 hours after fertilization, viewed in optical section (450X).

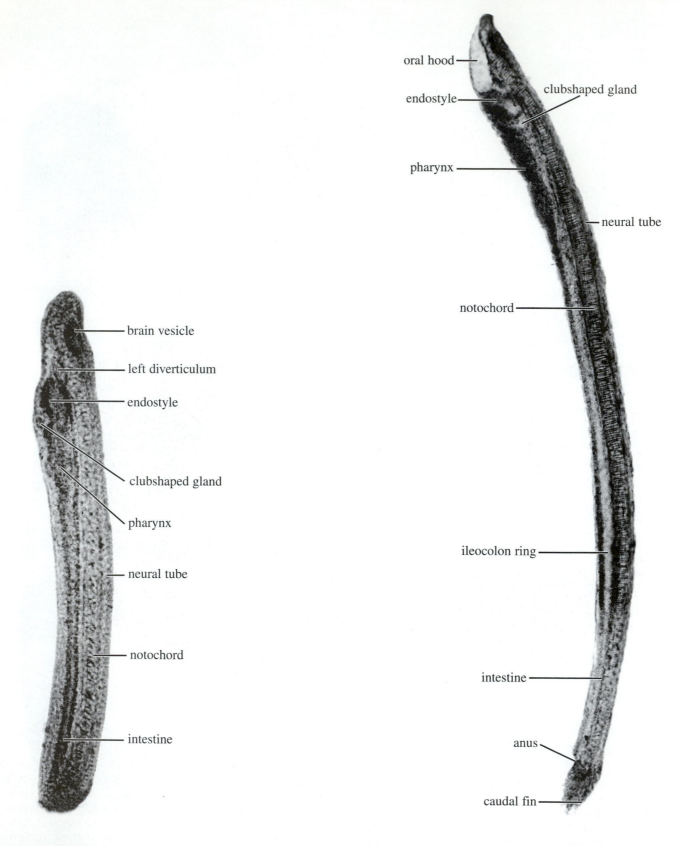

Figure 5.17

Amphioxus larva about 48 hours after fertilization; viewed in optical section (450X).

Figure 5.18

Amphioxus larva about 96 hours after fertilization; viewed in optical section (450X).

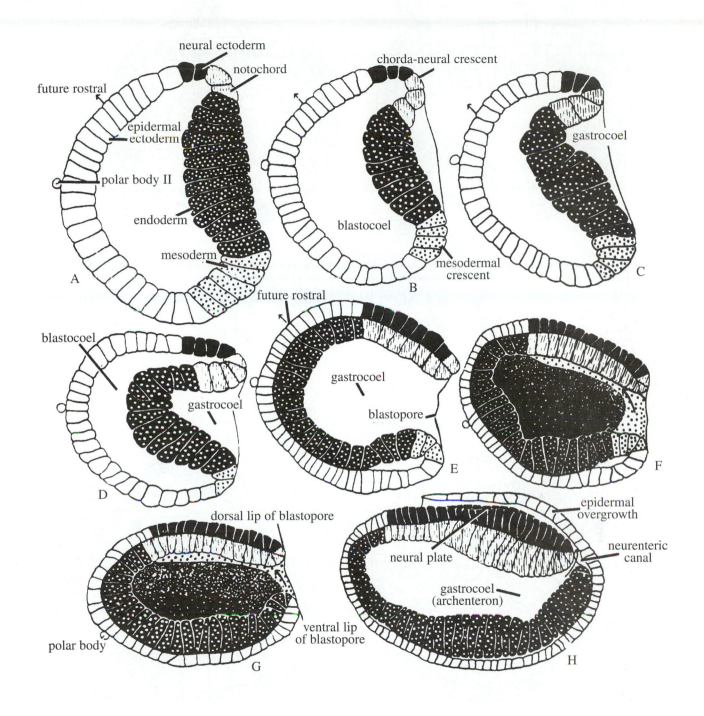

Figure 5.19

Gastrulation in *Amphioxus* showing the position of the organ-forming areas. A, beginning gastrulation; animal hemisphere of epidermal and neural ectoderm; vegetal hemisphere of notochord, endoderm and mesoderm is beginning to invaginate; B, the dorsal crescent of prospective notochord (chorda-neural crescent) is in the dorsal lip of blastopore, and the ventral crescent of mesoderm is in the ventral and lateral lips of the blastopore; C, D, invagination continues; E, gastrulation is almost completed and the blastocoel is nearly eliminated, bringing the ectoderm into broad contact with the endoderm of the archenteron; the neural area is underlain by notochord; F, G, as the embryo elongates and the blastopore constricts, the mesoderm splits and migrates rostrally along either side of the notochord; H, gastrulation is completed; the neural ectoderm thickens to form the neural plate and the neural folds; the ectoderm from the ventral lip overgrows the neural organ, forming the neurenteric canal.

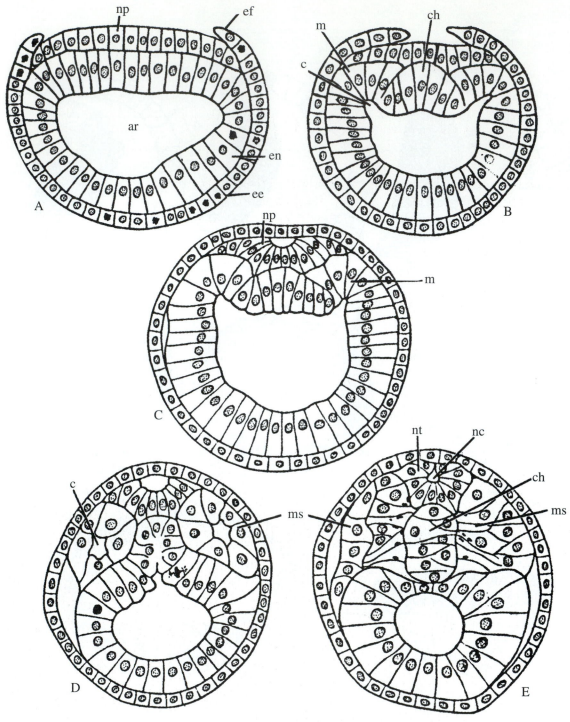

Figure 5.20

Transverse sections through *Amphioxus* embryos showing early organ formation. A, formation of neural plate and ectodermal folds from dorsal ectoderm; B, somites and notochord beginning to evaginate from archenteron; neural plate is becoming covered by advancing neural folds; C, evagination of somites proceeds, forming hollow vesicles; the neural plate folds longitudinally into the neural groove; D, first somites are now separated from the archenteron and possess an entrocoel cavity; E, section through the middle somites of a 9-somite larva; the notochord has separated from the archenteron; the dorsal edges of the archenteron are fused, and the neural groove has closed into a neural tube; the medial wall of the somites (myotome) is forming muscle fibrillae. ar, archenteron (gastrocoel); c, enterocoel; ch, notochord; ec, ectoderm; ef, ectodermal fold; en, endodermal tube; g, gut cavity; m, mesoderm; ms, mesodermal somites; nc, neural canal; np, neural plate; nt, neural tube.

Amphibian Development

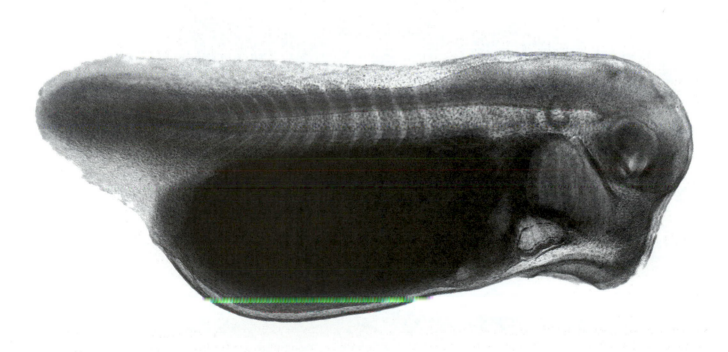

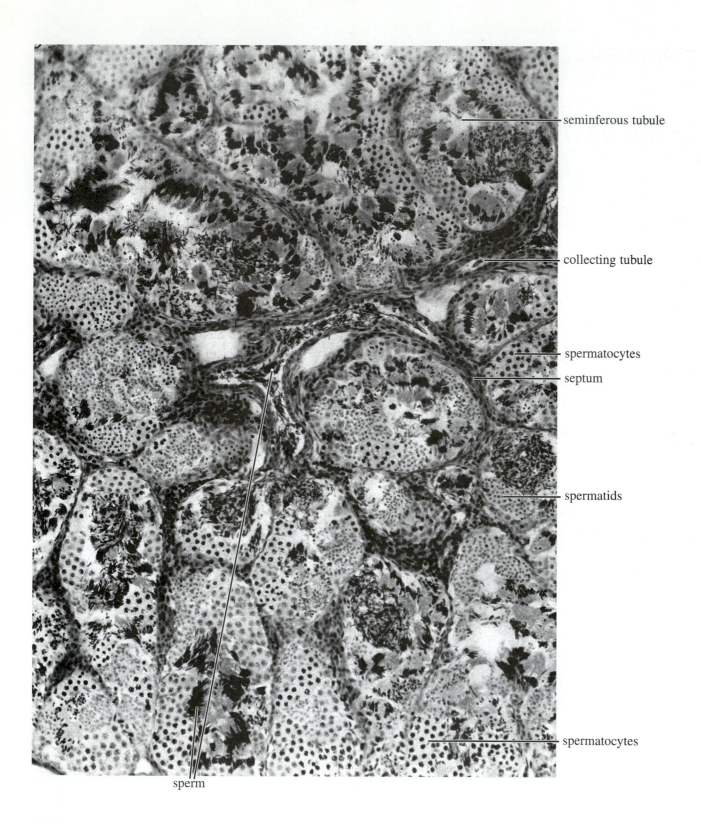

seminferous tubule

collecting tubule

spermatocytes

septum

spermatids

spermatocytes

sperm

Figure 6.1

Frog testis, section (225X).

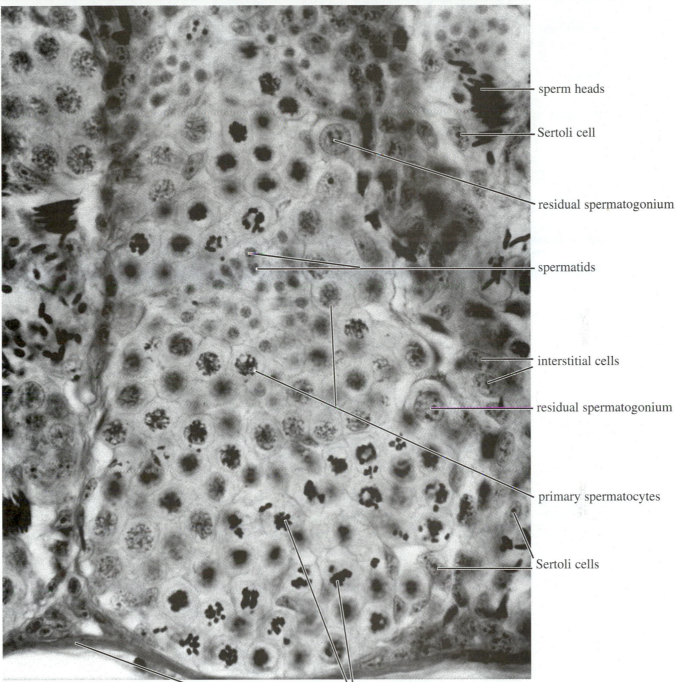

sperm heads

Sertoli cell

residual spermatogonium

spermatids

interstitial cells

residual spermatogonium

primary spermatocytes

Sertoli cells

tunica albuginea   maturation division 1, metaphase

Figure 6.2

Frog testis, section (725X).

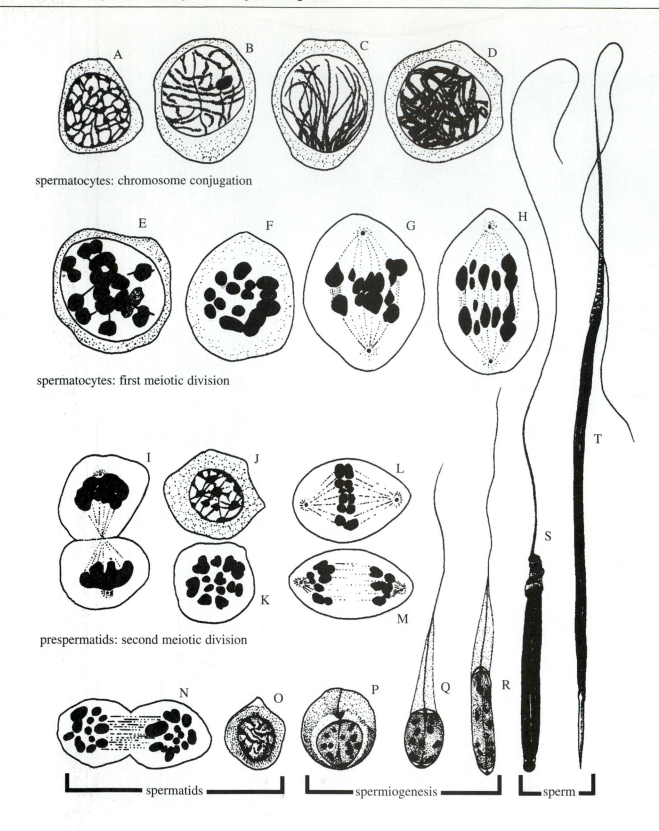

spermatocytes: chromosome conjugation

spermatocytes: first meiotic division

prespermatids: second meiotic division

spermatids — spermiogenesis — sperm

Figure 6.3

Spermatogenesis in the frog: Maturation phase. Drawings of single cells arranged in a progressive series, from spermatogonium just transforming into a spermatocyte (A), to mature sperm; the short and blut sylvatica type (S) is more common than is the pointed temporaria type (T) (1900X).

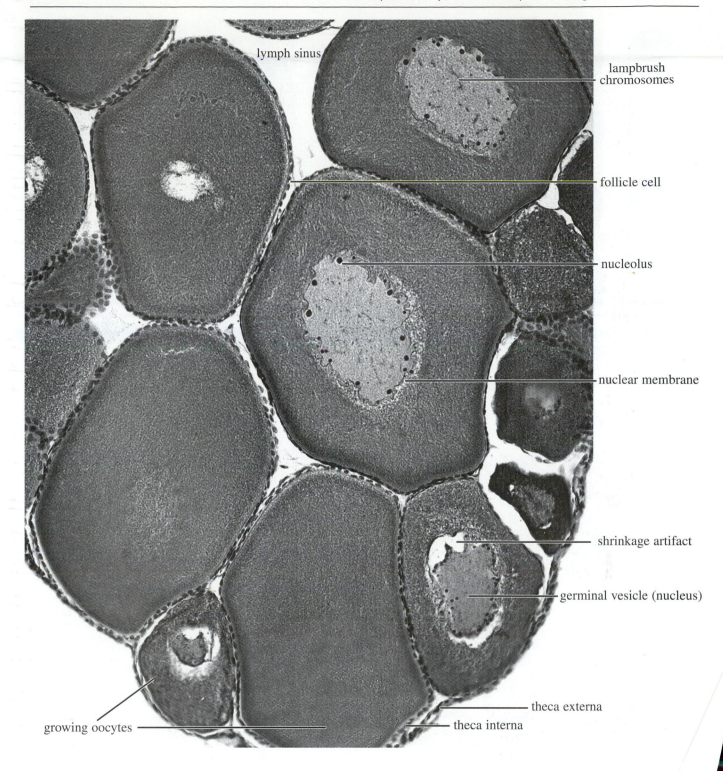

lymph sinus

lampbrush chromosomes

follicle cell

nucleolus

nuclear membrane

shrinkage artifact

germinal vesicle (nucleus)

theca externa

theca interna

growing oocytes

Figure 6.4

Frog ovary, section showing growing oocytes (180X).

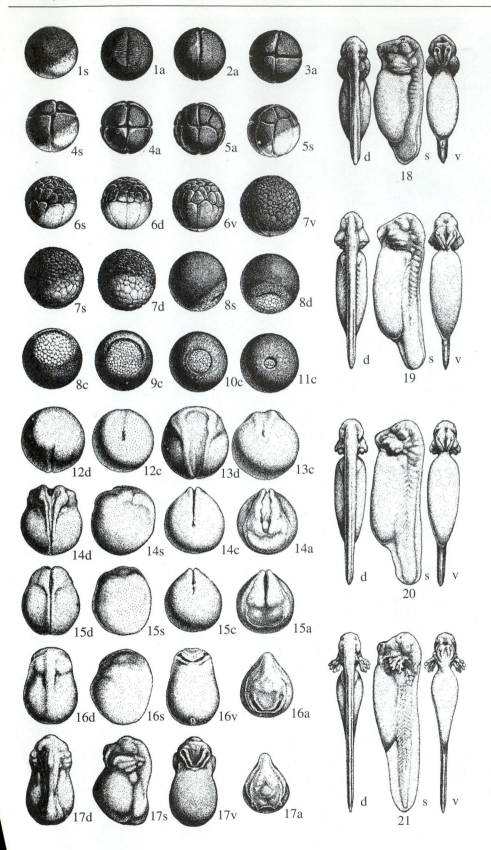

Figure 6.5

Witschi stages 1-21 of frog development *(Rana pipiens)*. Numbers 1-21 designate developmental stages (see Table 1). a, view from animal pole (frontal view); c, caudal (blastoporal) view; d, dorsal view; s, left lateral view; v, ventral view (6.5X).

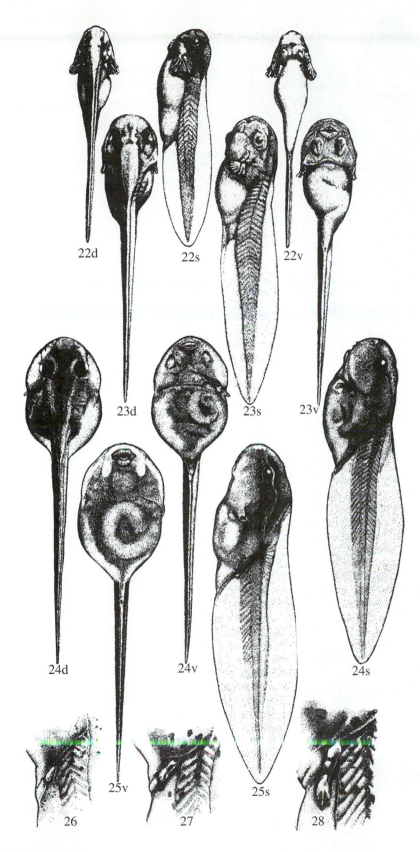

Figure 6.6

Witschi stages 22-28 of frog development *(Rana pipiens)*. Numbers 22-28 designate developmental stages (see Table 1). d, dorsal view; s, left lateral view; v, ventral view (Stages 22-25, 6.5X; Stages 26-28, 20X).

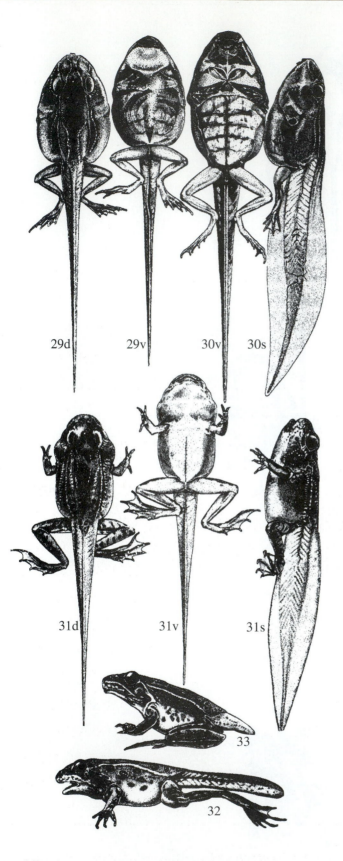

Figure 6.7

Witschi stages 29-33 of frog development *(Rana pipiens)*. Numbers 29-33 designate developmental stages (see Table 1). d, dorsal view; s, left lateral view; v, ventral view (8X).

Table 1

Frog Development Stages, *Rana pipens* (see Figures 6.5-6.7)

Witschi Stage Numbers*	Approximate Lengths in mm**	Key Criteria Defining Stages**
1	1.7	fertilized egg
2		two cells
3		four cells
4		eight cells
5		sixteen cells
6		early blastula
7		late blastula
8		early gastula
9		middle gastula
10		yolk plug
11		late gastrula
12		neural plate
13		neural folds
14		early neural groove
15		late neural groove
16	2.5-2.7	early neural tube
17	2.8-3.0	early tail bud
18	4	tail bud
19	5	gill buds
20	6	hatching, gill circulation
21	7	mouth open
22	8	tail fin circulation
23	9	opercular fold
24	10	right operculum closed
25	11	operculum complete
26-33		metamorphosis

*Similar to but not identical with Shumway stage numbers.
**Based on Shumway, W., Stages in the normal development of *Rana pipiens*. I. External form. *Anatomical Record* 78:139 (1940).

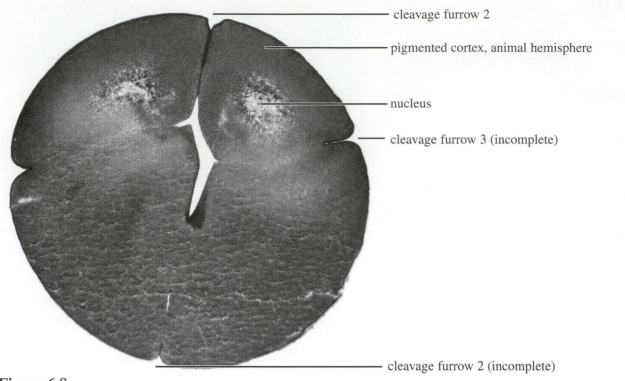

cleavage furrow 2

pigmented cortex, animal hemisphere

nucleus

cleavage furrow 3 (incomplete)

cleavage furrow 2 (incomplete)

Figure 6.8

Frog embryo, early cleavage, 8 cells (stage 4), median section (65X).

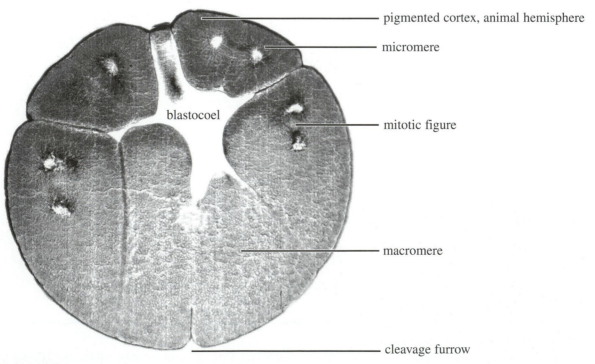

pigmented cortex, animal hemisphere

micromere

blastocoel

mitotic figure

macromere

cleavage furrow

Figure 6.9

Frog embryo, early cleavage, 32 cells (stage 6), median section (65X).

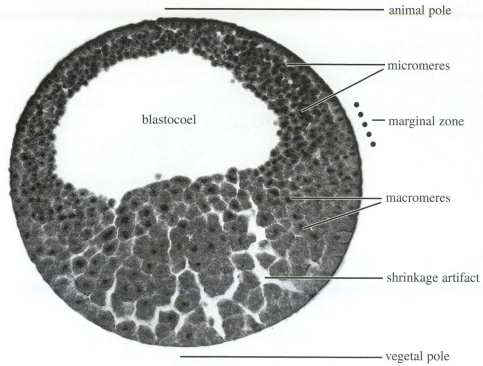

animal pole

micromeres

marginal zone

blastocoel

macromeres

shrinkage artifact

vegetal pole

Figure 6.10

Frog embryo, late cleavage, blastula (stage 7), median section (65X).

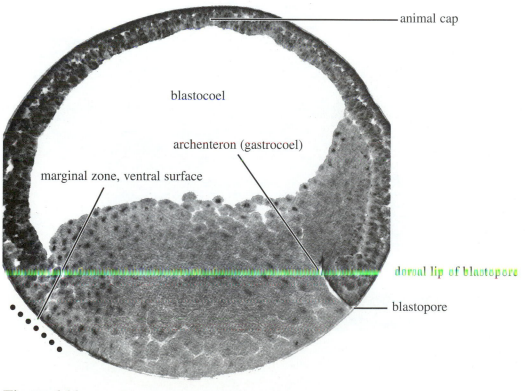

animal cap

blastocoel

archenteron (gastrocoel)

marginal zone, ventral surface

dorsal lip of blastopore

blastopore

Figure 6.11

Frog embryo, early gastula (stage 8), sagittal section (65X).

fertilization membrane   outer layer of ectoderm   inner layer of ectoderm   archenteron roof (chordamesoderm)

dorsal lip of blastopore

archenteron (gastrocoel)

blastopore

blastocoel   yolk plug

blastopore

ventral lip of blastopore

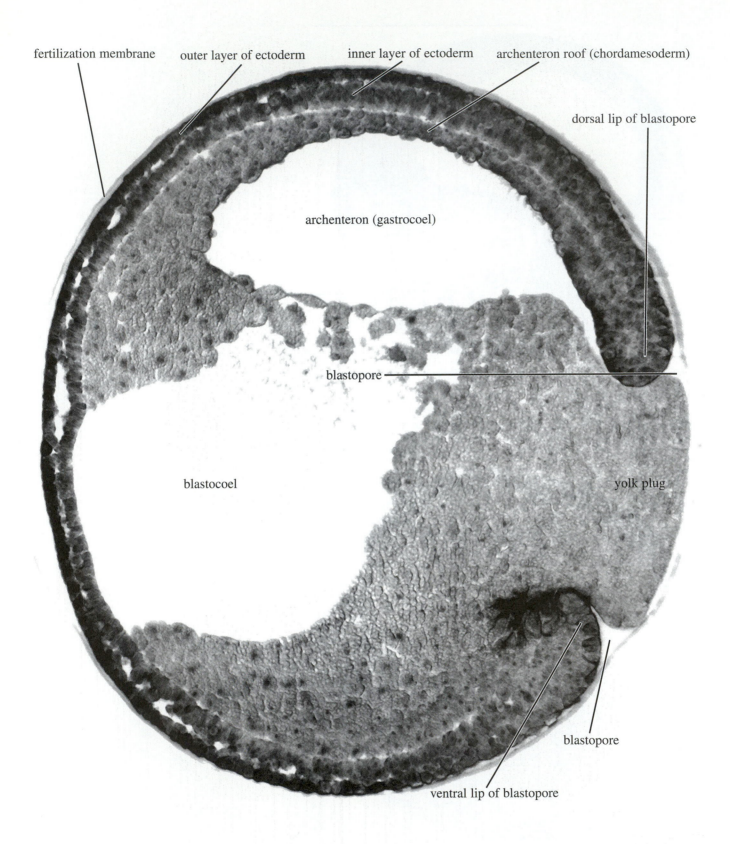

Figure 6.12

Frog embryo, late gastrula (stage 10), sagittal section (100X).

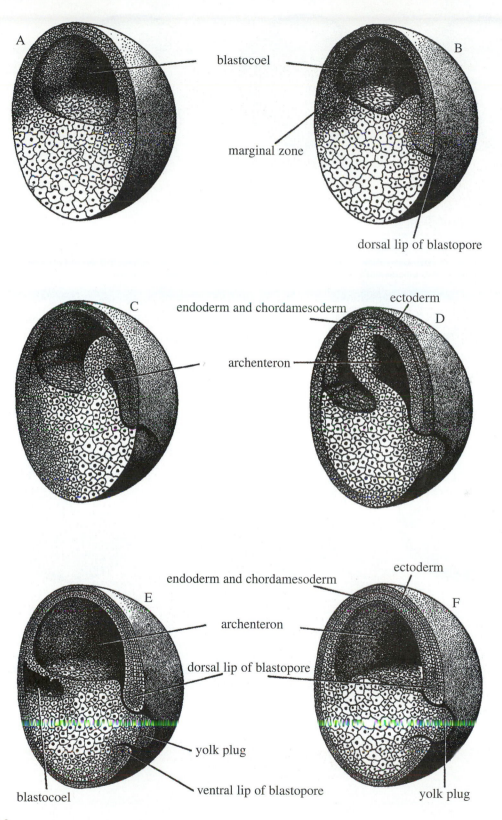

Figure 6.13

The blastula (A) of the frog and its transformation into the gastrula (B-F). B, beginning of gastrulation; C-E, elimination of the blastocoel or segmentation cavity by the gastrocoel or archenteron; F, nearly completed gastrula with mesoderm and endoderm beneath the ectoderm.

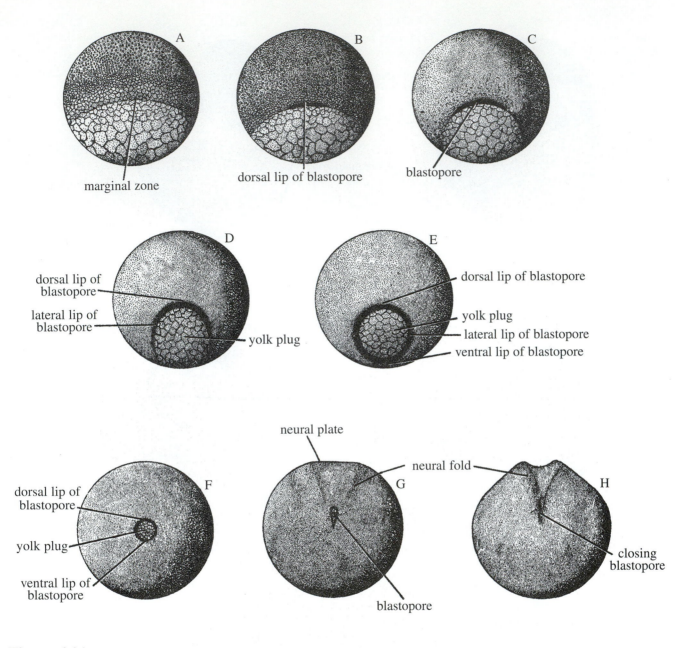

Figure 6.14

Gastrulation in the frog embryo as seen from the caudal or blastoporal point of view. Stages A-E are equivalent to those contained in Figure 6.13. Stages F and G are almost the same as those of D and E of Figure 6.15.

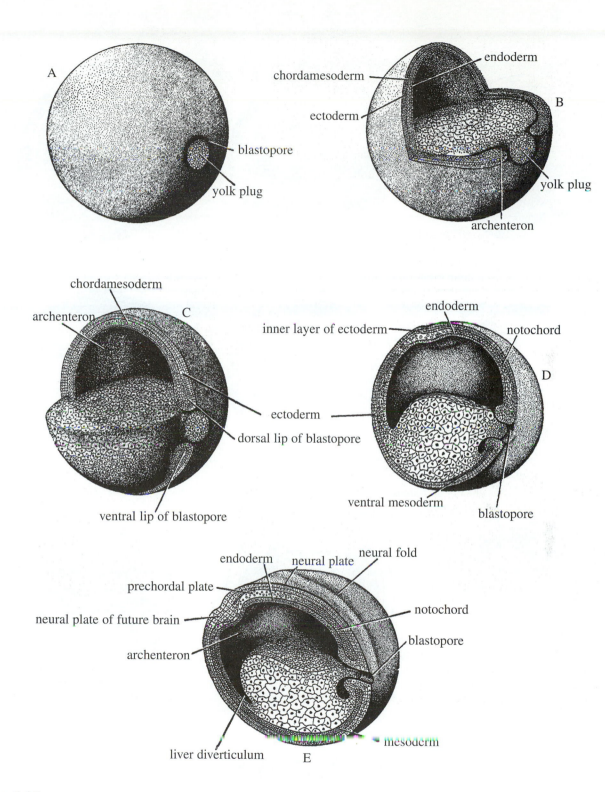

Figure 6.15

The yolk-plug gastrula (A) of the frog and its transformation into the late gastrula (B-E). A, external appearance of the yolk-plug gastrula; B, C, different partial sections of the yolk-plug gastrula; D, E, late gastrula stages; the blastopore is becoming smaller, the yolk plug is withdrawing and the embryo is elongating in the rostrocaudal direction.

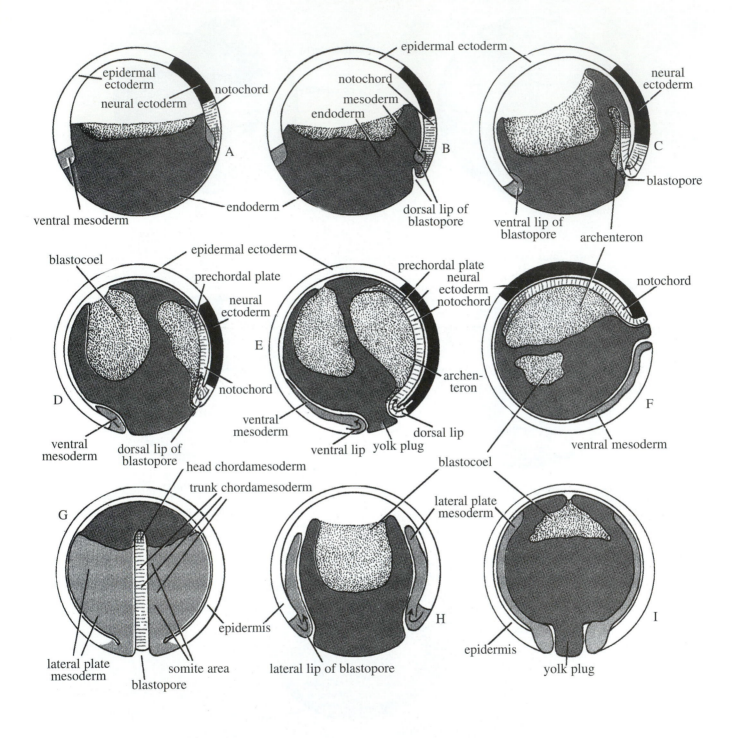

Figure 6.16

Migration of the presumptive organ-forming areas of the blastula during gastrulation in the amphibia (with reference particularly to the frog). A, late blastula; sagittal section through the midplane of the future embryo; B-F, epiboly and emboly; in epiboly, the black (neural) and white (epidermal) areas become extended and gradually envelop the inwardly moving notochord, endoderm and mesoderm. In emboly, the notochord, endoderm and mesoderm move inward; G, late gastrula stage; the neural ectoderm and the upper portion of the epidermal ectoderm have been removed to show the relationships of the middle germ layer of chordamesoderm; H, horizontal section of the mid-gastrula stage; I, late gastrula stage; horizontal section showing the yolk plug, mesoderm and obliteration of the blastocoel by endoderm.

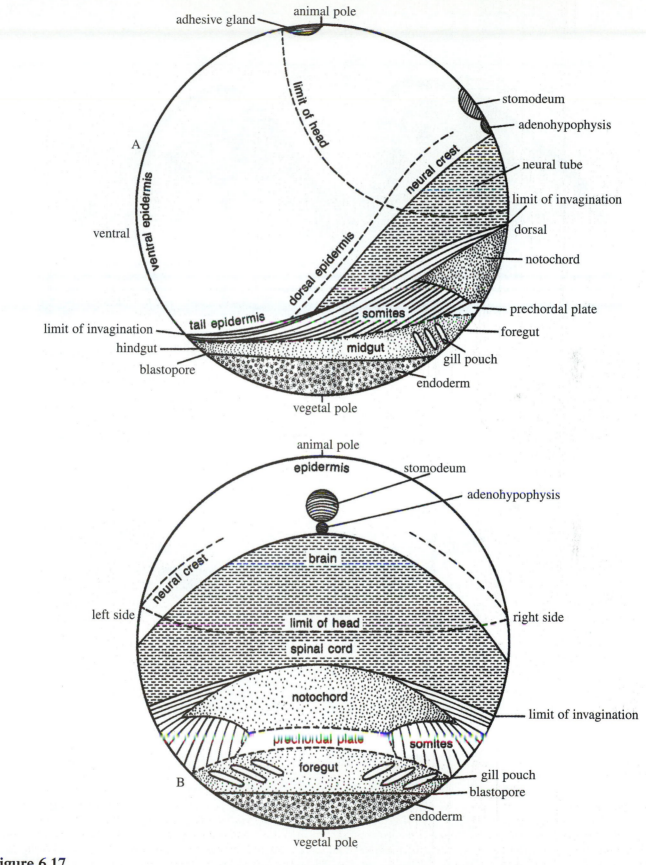

Figure 6.17

Maps of the presumptive regions of the very young gastrula of an anuran. A, side view; B, dorsal view.

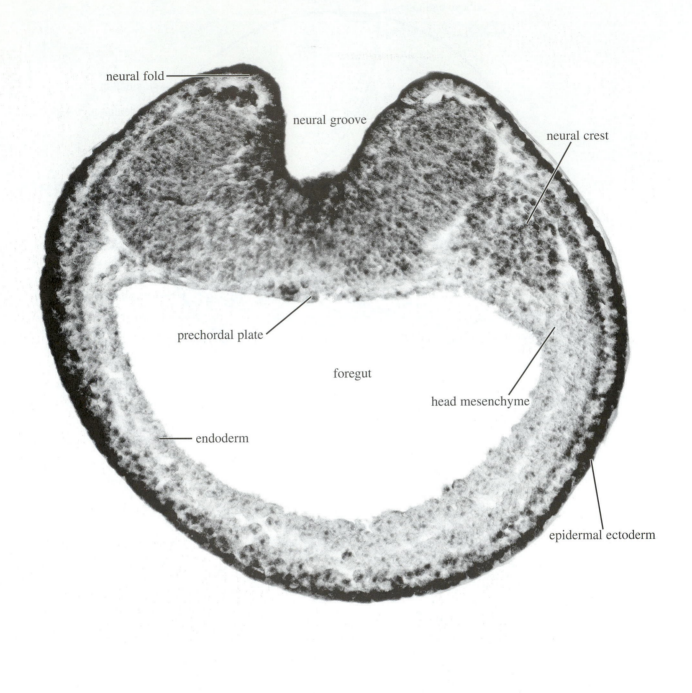

neural fold

neural groove

neural crest

prechordal plate

foregut

head mesenchyme

endoderm

epidermal ectoderm

Figure 6.18

Frog embryo, neural fold/groove stage (stages 12-15), transverse section through the head (100X).

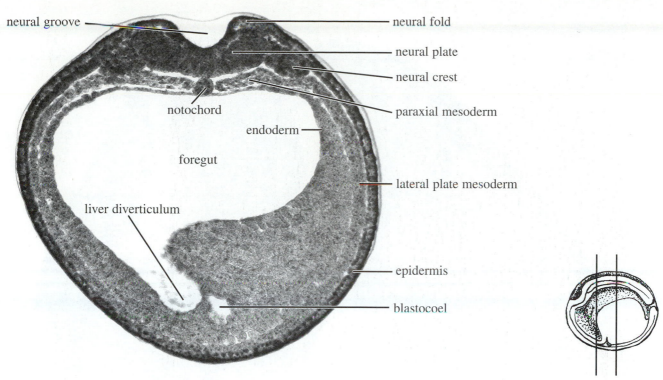

neural groove

neural fold

neural plate

neural crest

notochord

paraxial mesoderm

endoderm

foregut

lateral plate mesoderm

liver diverticulum

epidermis

blastocoel

Figure 6.19

Frog embryo, neural fold/groove stage (stages 12-15), transverse section through the foregut (65X).

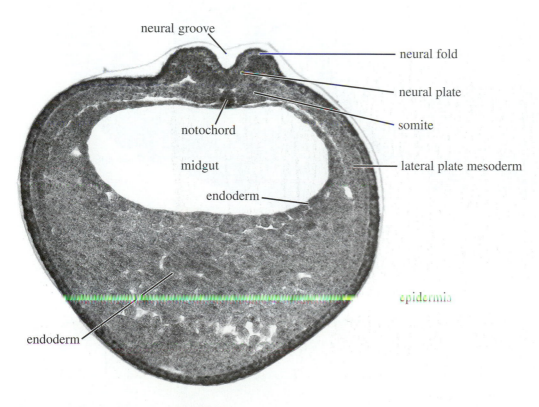

neural groove

neural fold

neural plate

notochord

somite

midgut

lateral plate mesoderm

endoderm

epidermis

endoderm

Figure 6.20

Frog embryo, neural fold/groove stage (stages 12-15), transverse section through the midgut (65X).

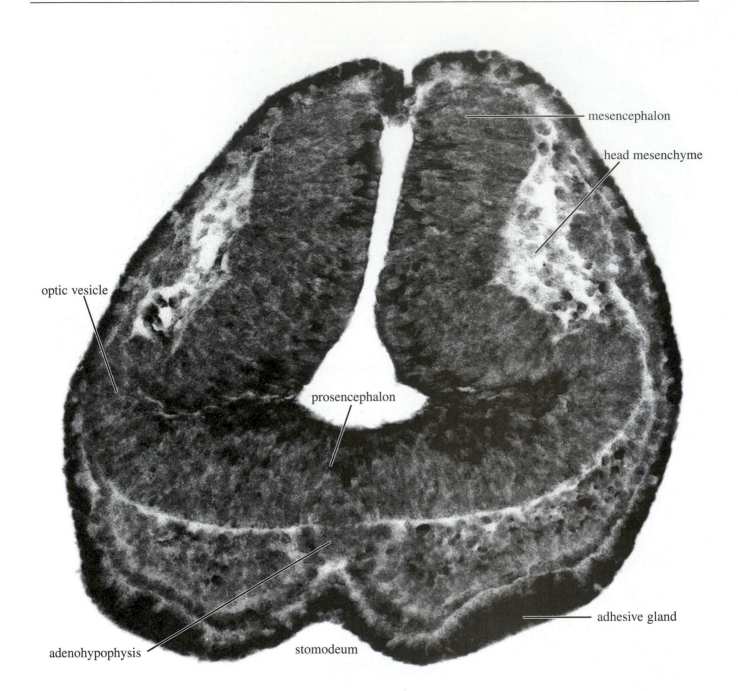

mesencephalon

head mesenchyme

optic vesicle

prosencephalon

adhesive gland

adenohypophysis

stomodeum

Figure 6.21

Frog embryo, neural tube stage (stage 16), transverse section through the optic vesicles (100X).

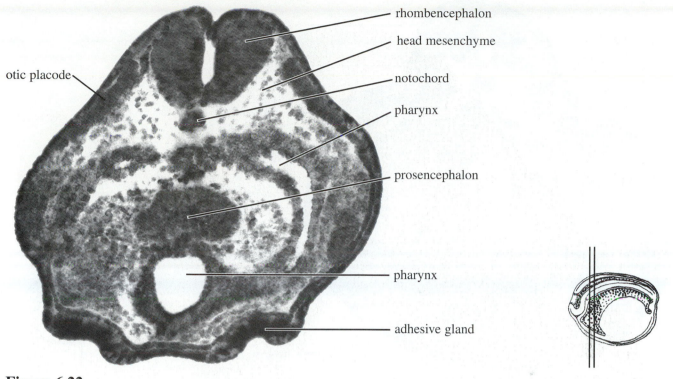

rhombencephalon

head mesenchyme

otic placode

notochord

pharynx

prosencephalon

pharynx

adhesive gland

Figure 6.22

Frog embryo, neural tube stage (stage 16), transverse section through the otic placodes (65X).

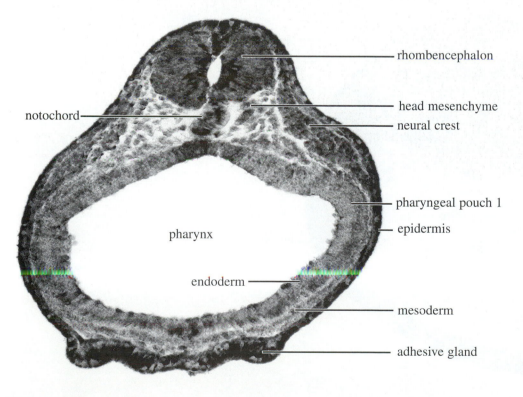

rhombencephalon

notochord

head mesenchyme

neural crest

pharyngeal pouch 1

epidermis

pharynx

endoderm

mesoderm

adhesive gland

Figure 6.23

Frog embryo, neural tube stage (stage 16), transverse section through the pharynx (65X).

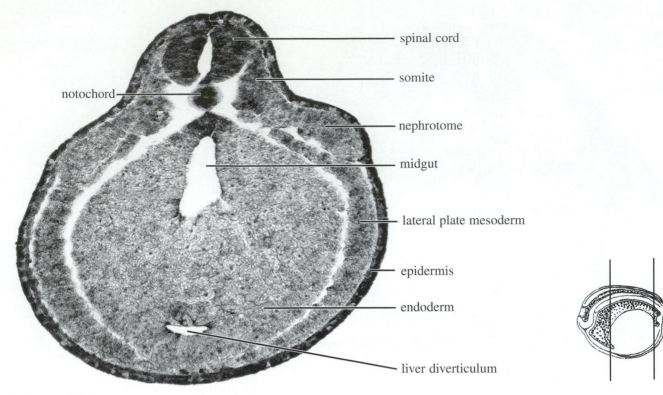

- spinal cord
- somite
- notochord
- nephrotome
- midgut
- lateral plate mesoderm
- epidermis
- endoderm
- liver diverticulum

Figure 6.24

Frog embryo, neural tube stage (stage 16), transverse section through the nephrotome (65X).

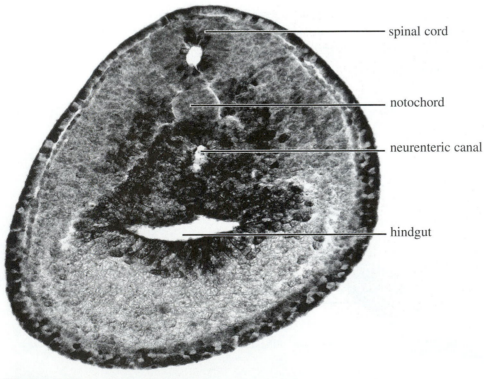

- spinal cord
- notochord
- neurenteric canal
- hindgut

Figure 6.25

Frog embryo, neural tube stage (stage 16), transverse section through the hindgut (65X).

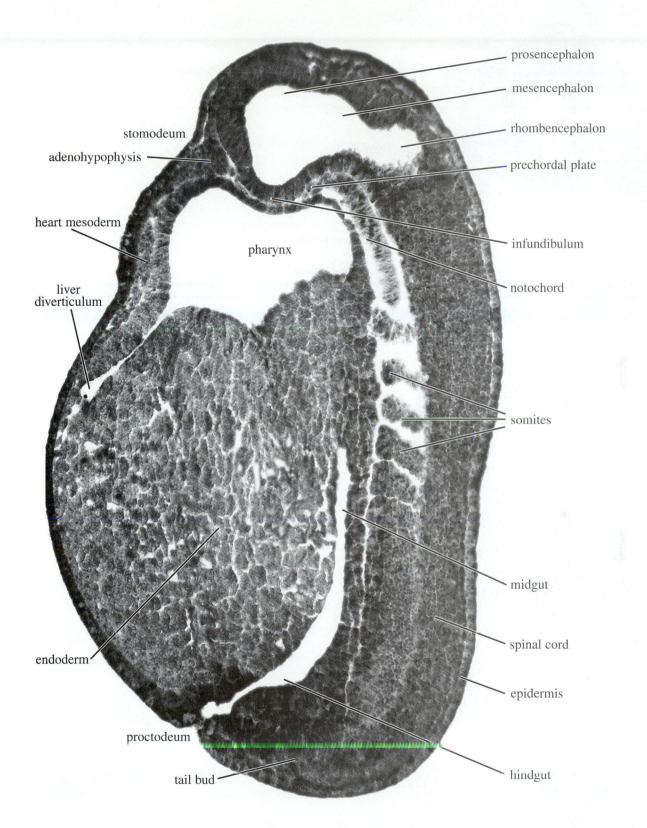

prosencephalon

mesencephalon

rhombencephalon

stomodeum

adenohypophysis

prechordal plate

heart mesoderm

infundibulum

pharynx

notochord

liver
diverticulum

somites

midgut

spinal cord

endoderm

epidermis

proctodeum

tail bud

hindgut

Figure 6.26

Frog embryo, tail bud stage (stage 17), sagittal section (80X).

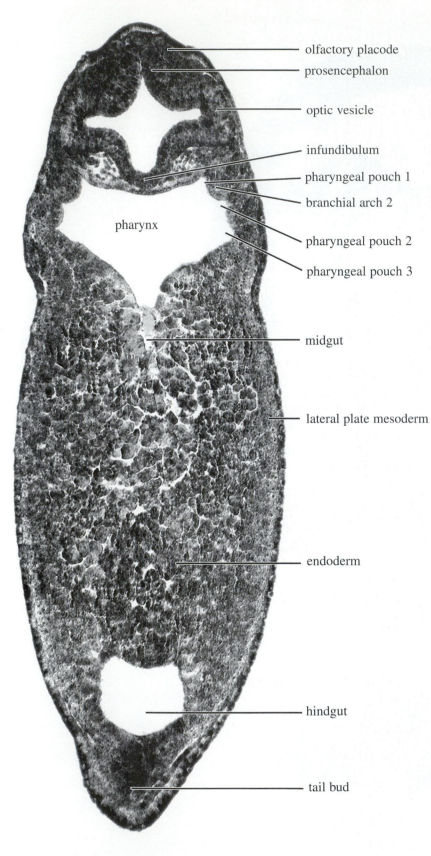

olfactory placode

prosencephalon

optic vesicle

infundibulum

pharyngeal pouch 1

branchial arch 2

pharynx

pharyngeal pouch 2

pharyngeal pouch 3

midgut

lateral plate mesoderm

endoderm

hindgut

tail bud

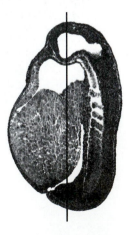

Figure 6.27

Frog embryo, tail bud stage (stage 17), frontal section through the pharynx (80X).

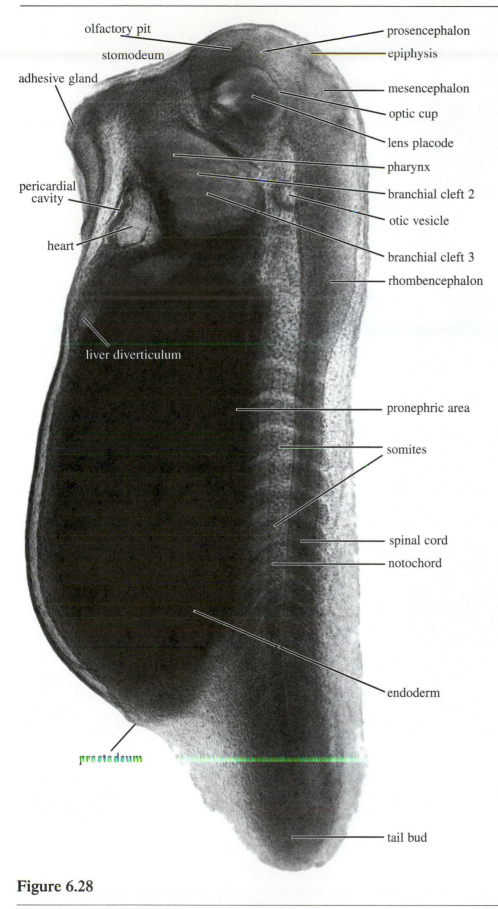

olfactory pit

stomodeum

adhesive gland

pericardial cavity

heart

liver diverticulum

proctodeum

prosencephalon

epiphysis

mesencephalon

optic cup

lens placode

pharynx

branchial cleft 2

otic vesicle

branchial cleft 3

rhombencephalon

pronephric area

somites

spinal cord

notochord

endoderm

tail bud

Figure 6.28

4-mm frog embryo (stage 18), whole mount (60X).

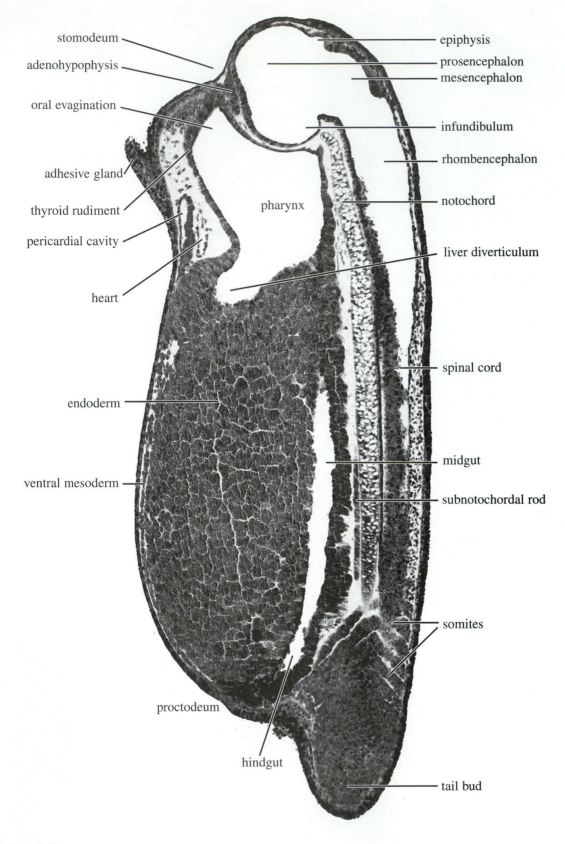

Figure 6.29

4-mm frog embryo (stage 18), sagittal section (60X).

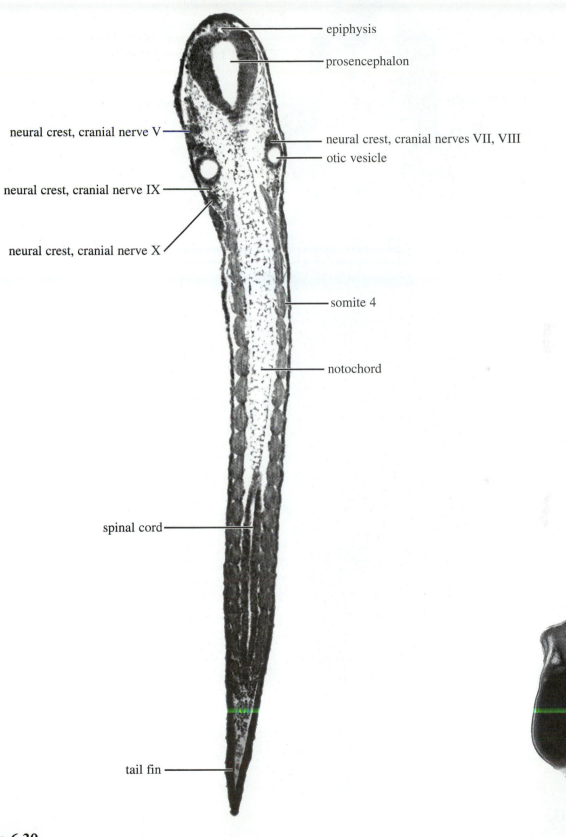

epiphysis

prosencephalon

neural crest, cranial nerve V

neural crest, cranial nerves VII, VIII

otic vesicle

neural crest, cranial nerve IX

neural crest, cranial nerve X

somite 4

notochord

spinal cord

tail fin

Figure 6.30

4-mm frog embryo (stage 18), frontal section through the otic vesicles (60X).

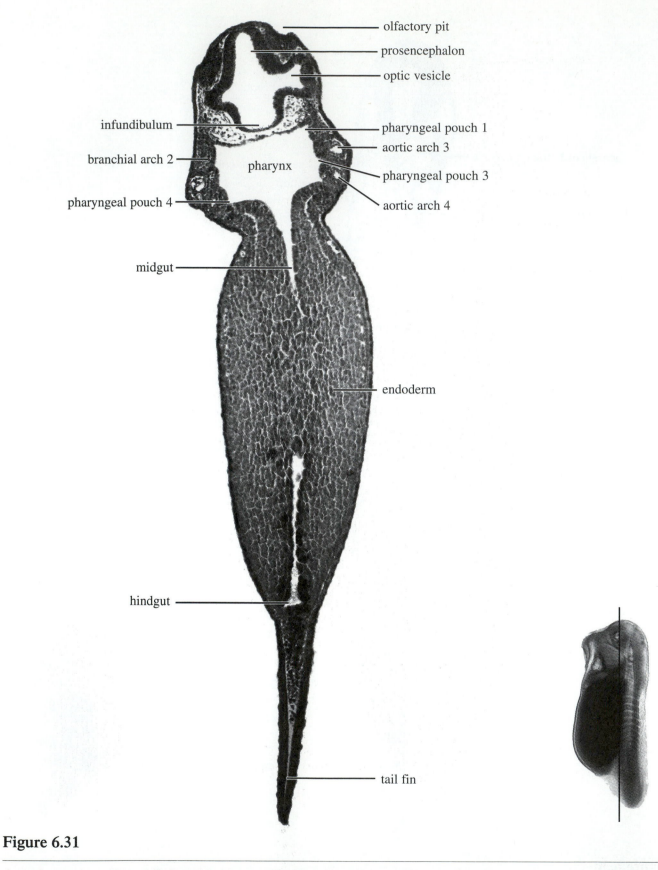

Figure 6.31

4-mm frog embryo (stage 18), frontal section through the optic vesicles (60X).

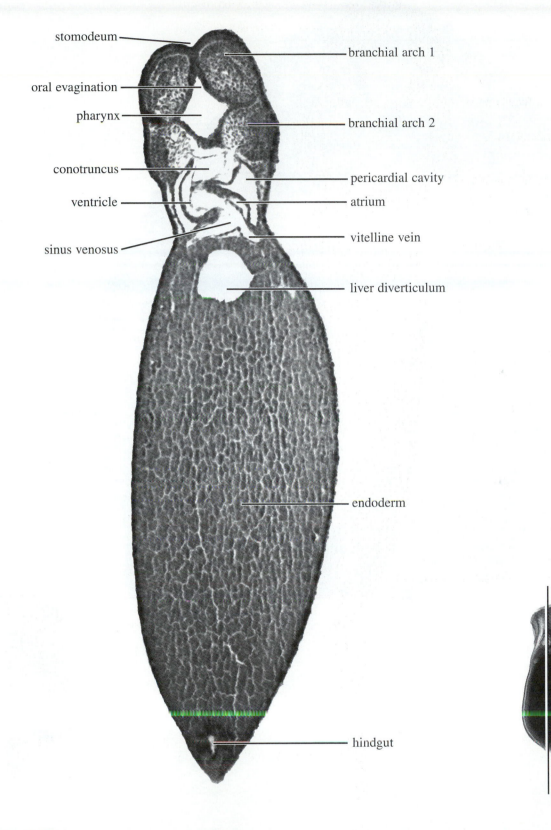

stomodeum

oral evagination

pharynx

conotruncus

ventricle

sinus venosus

branchial arch 1

branchial arch 2

pericardial cavity

atrium

vitelline vein

liver diverticulum

endoderm

hindgut

Figure 6.32

4-mm frog embryo (stage 18), frontal section through the heart (60X).

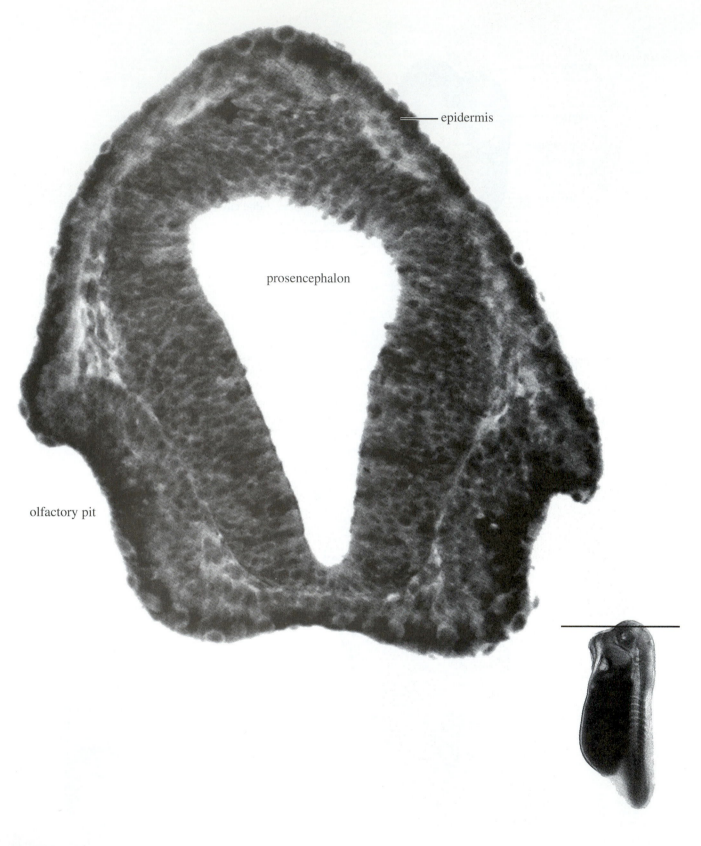

Figure 6.33

4-mm frog embryo (stage 18), transverse section through the olfactory pits (260X).

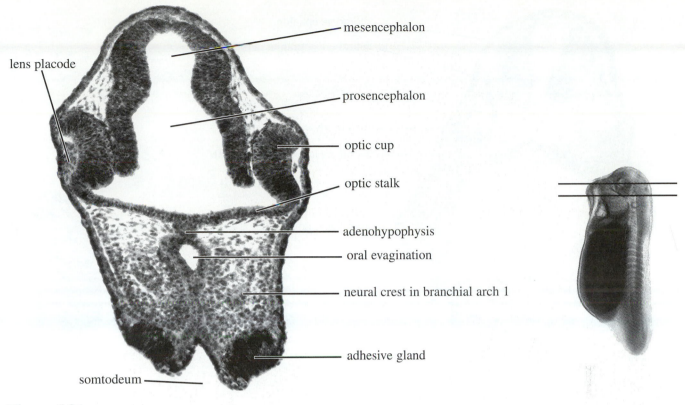

mesencephalon

lens placode

prosencephalon

optic cup

optic stalk

adenohypophysis

oral evagination

neural crest in branchial arch 1

adhesive gland

somtodeum

Figure 6.34

4-mm frog embryo (stage 18), transverse section through the optic cups (85X).

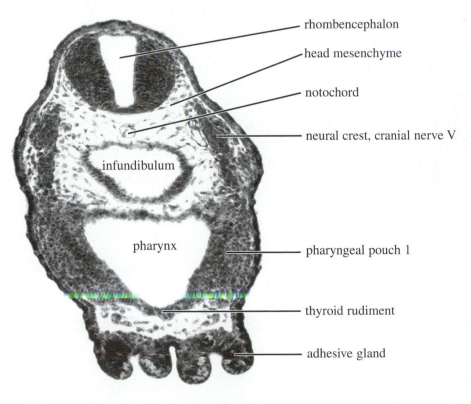

rhombencephalon

head mesenchyme

notochord

neural crest, cranial nerve V

infundibulum

pharynx

pharyngeal pouch 1

thyroid rudiment

adhesive gland

Figure 6.35

4-mm frog embryo (stage 18), transverse section through the rostral pharynx (85X).

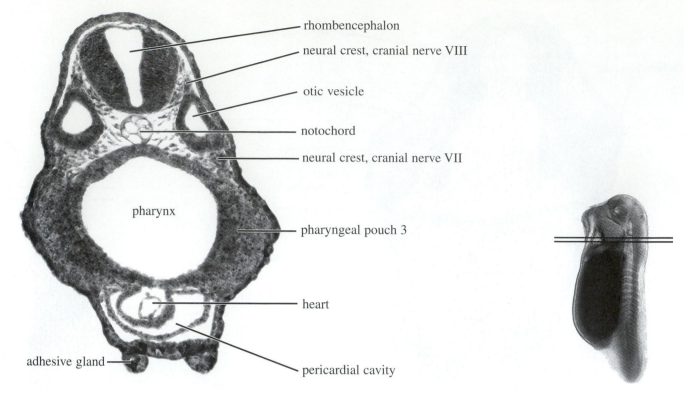

rhombencephalon

neural crest, cranial nerve VIII

otic vesicle

notochord

neural crest, cranial nerve VII

pharynx

pharyngeal pouch 3

heart

adhesive gland

pericardial cavity

Figure 6.36

4-mm frog embryo (stage 18), transverse section through the otic vesicles (85X).

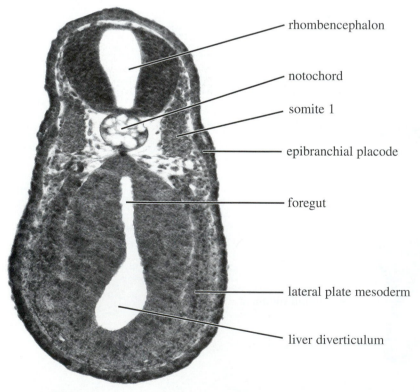

rhombencephalon

notochord

somite 1

epibranchial placode

foregut

lateral plate mesoderm

liver diverticulum

Figure 6.37

4-mm frog embryo (stage 18), transverse section through the liver diverticulum (85X).

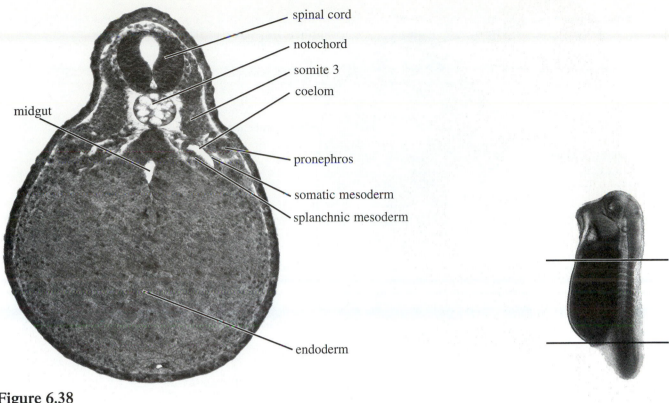

midgut

spinal cord

notochord

somite 3

coelom

pronephros

somatic mesoderm

splanchnic mesoderm

endoderm

Figure 6.38

4-mm frog embryo (stage 18), transverse section through the pronephros (85X).

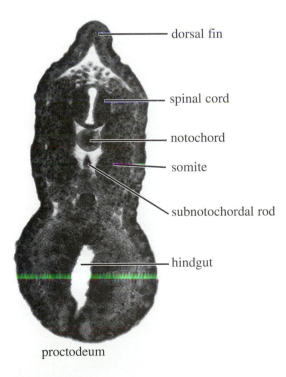

dorsal fin

spinal cord

notochord

somite

subnotochordal rod

hindgut

proctodeum

Figure 6.39

4-mm frog embryo (stage 18), transverse section through the hindgut (85X).

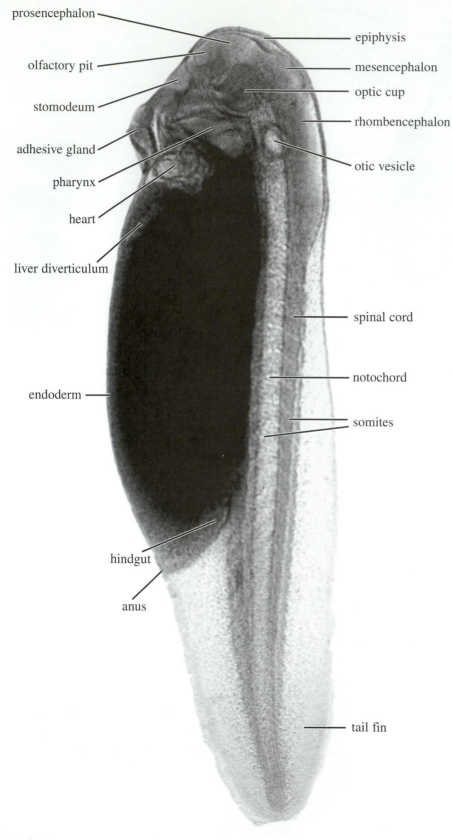

prosencephalon

epiphysis

olfactory pit

mesencephalon

stomodeum

optic cup

rhombencephalon

adhesive gland

otic vesicle

pharynx

heart

liver diverticulum

spinal cord

notochord

endoderm

somites

hindgut

anus

tail fin

Figure 6.40

6/7-mm frog tadpole (stages 20-21), whole mount (45X).

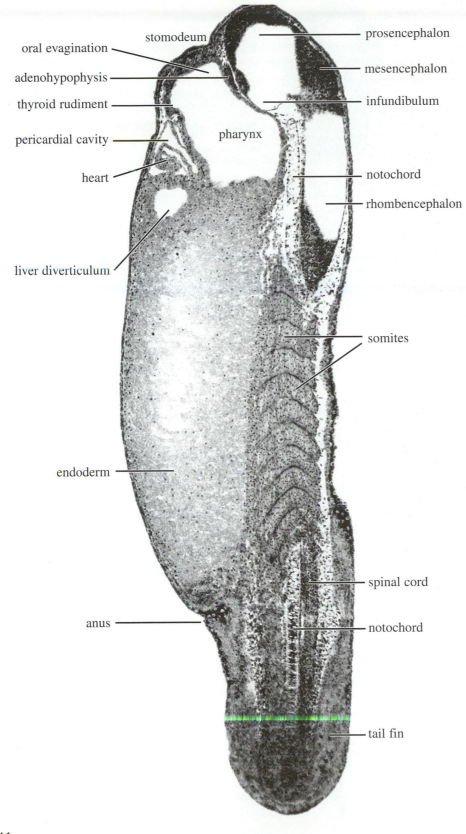

Figure 6.41

6/7-mm frog tadpole (stages 20-21), sagittal section (45X).

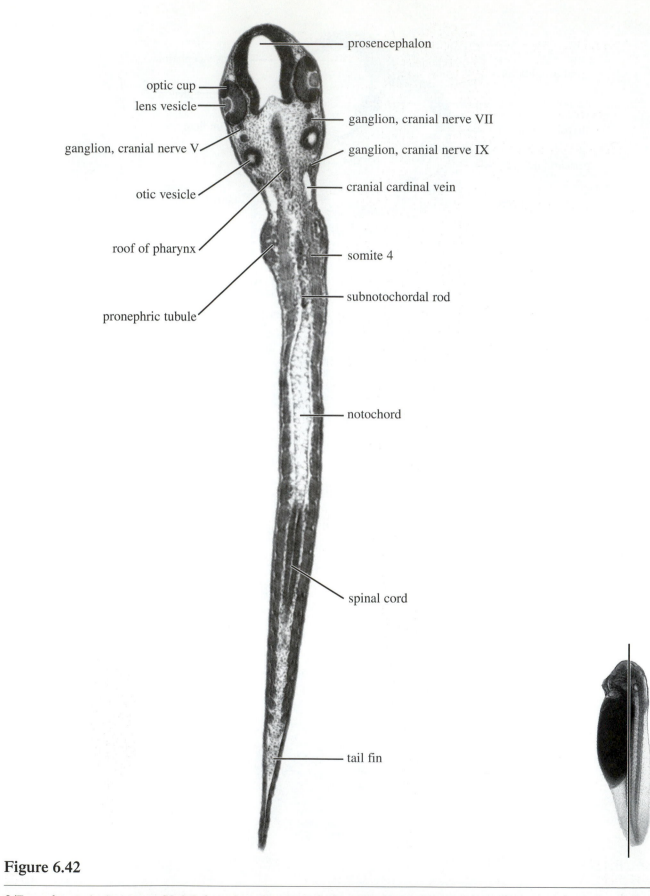

prosencephalon

optic cup

lens vesicle

ganglion, cranial nerve VII

ganglion, cranial nerve V

ganglion, cranial nerve IX

otic vesicle

cranial cardinal vein

roof of pharynx

somite 4

subnotochordal rod

pronephric tubule

notochord

spinal cord

tail fin

Figure 6.42

6/7-mm frog tadpole (stages 20-21), frontal section through the optic cups (45X).

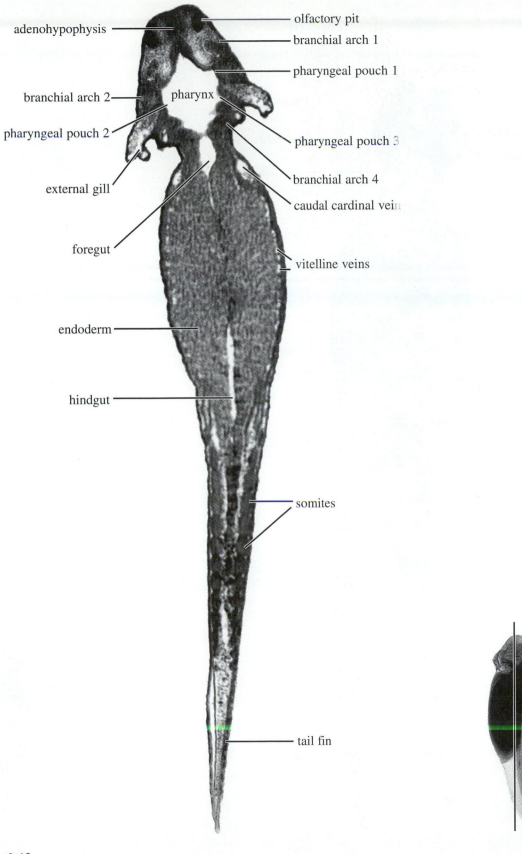

adenohypophysis

olfactory pit

branchial arch 1

pharyngeal pouch 1

branchial arch 2

pharynx

pharyngeal pouch 2

pharyngeal pouch 3

external gill

branchial arch 4

foregut

caudal cardinal vein

vitelline veins

endoderm

hindgut

somites

tail fin

Figure 6.43

6/7-mm frog tadpole (stages 20-21), frontal section through the pharynx (45X).

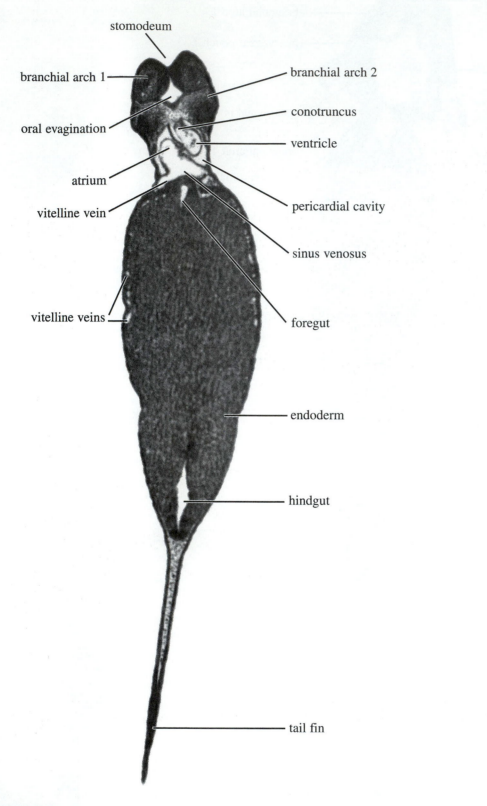

stomodeum

branchial arch 1

branchial arch 2

oral evagination

conotruncus

ventricle

atrium

vitelline vein

pericardial cavity

sinus venosus

vitelline veins

foregut

endoderm

hindgut

tail fin

Figure 6.44

6/7-mm frog tadpole (stages 20-21), frontal section through the heart (45X).

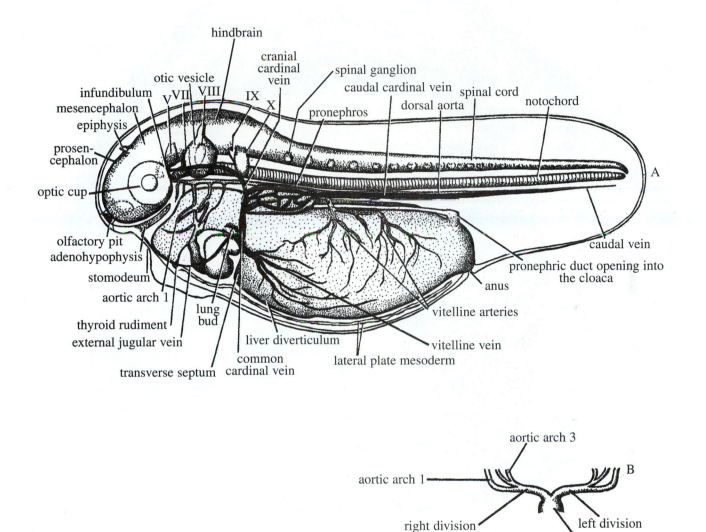

Figure 6.45

A, drawing of the 6/7-mm frog tadpole; B, drawing of the aortic arches of the 6/7-mm frog tadpole.

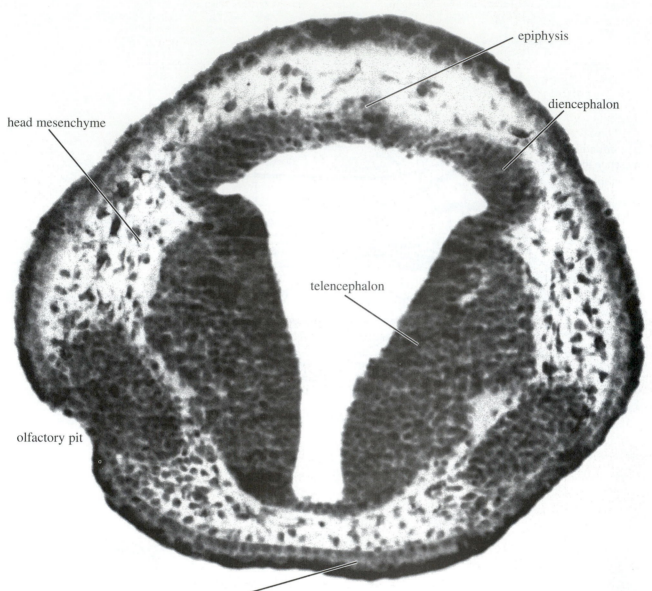

epiphysis

diencephalon

head mesenchyme

telencephalon

olfactory pit

epidermal ectoderm

Figure 6.46

6/7-mm frog tadpole (stages 20-21), transverse section through the olfactory pits (225X).

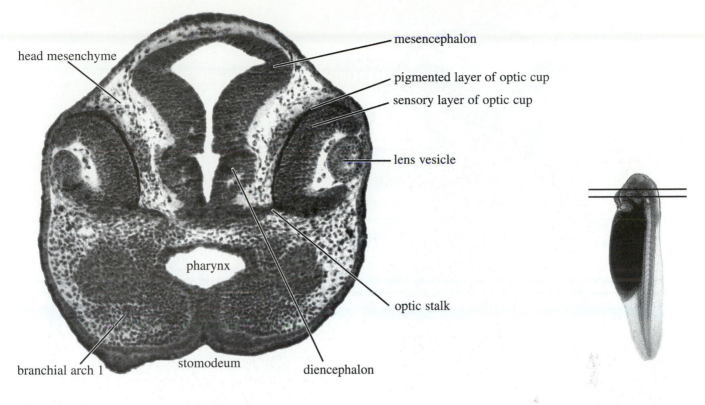

Figure 6.47

6/7-mm frog tadpole (stages 20-21), transverse section through the optic cups (100X).

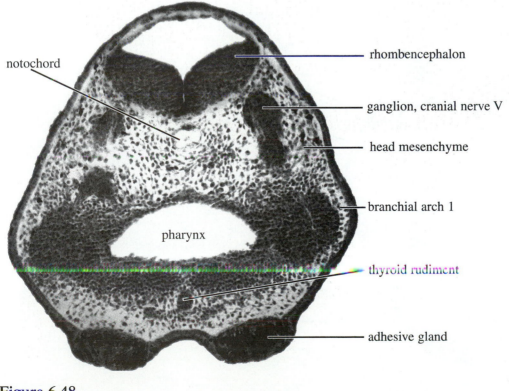

Figure 6.48

6/7-mm frog tadpole (stages 20-21), transverse section through the thyroid rudiment (100X).

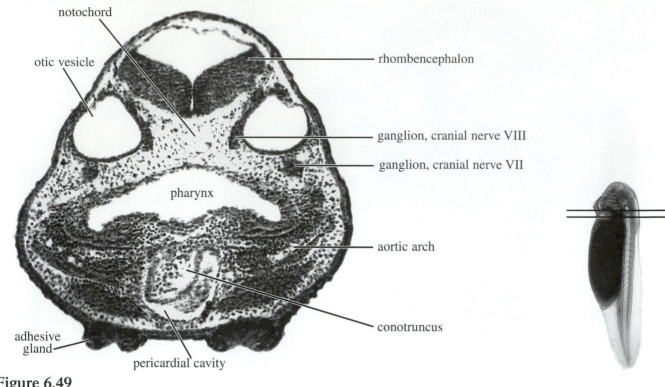

notochord

otic vesicle

rhombencephalon

ganglion, cranial nerve VIII

ganglion, cranial nerve VII

pharynx

aortic arch

conotruncus

adhesive
gland

pericardial cavity

Figure 6.49

6/7-mm frog tadpole (stages 20-21), transverse section through the otic vesicles (100X).

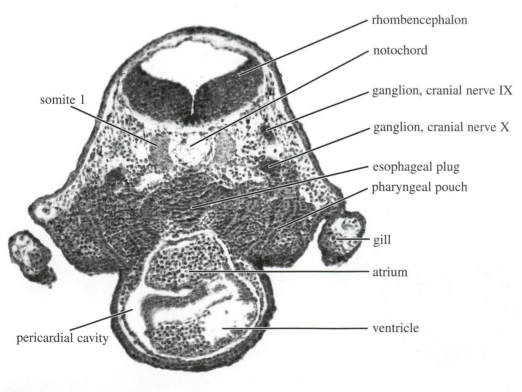

rhombencephalon

notochord

ganglion, cranial nerve IX

ganglion, cranial nerve X

somite 1

esophageal plug

pharyngeal pouch

gill

atrium

ventricle

pericardial cavity

Figure 6.50

6/7-mm frog tadpole (stages 20-21), transverse section through the heart (100X).

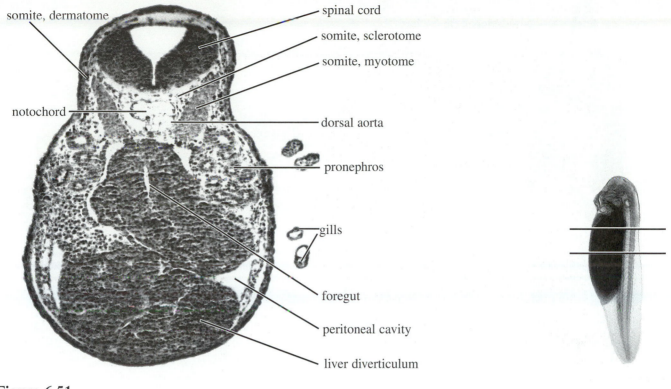

Figure 6.51

6/7-mm frog tadpole (stages 20-21), transverse section through the pronephros (100X).

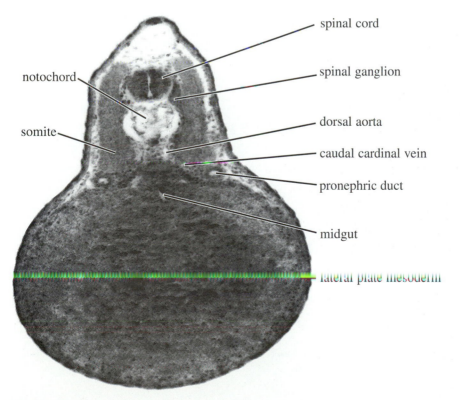

Figure 6.52

6/7-mm frog tadpole (stages 20-21), transverse section through the midgut (100X).

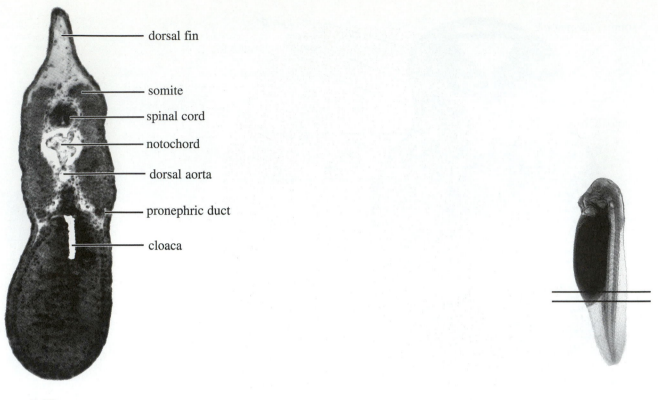

— dorsal fin

— somite

— spinal cord

— notochord

— dorsal aorta

— pronephric duct

— cloaca

Figure 6.53

6/7-mm frog tadpole (stages 20-21), transverse section through the cloaca (100X).

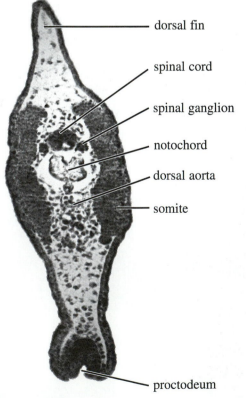

— dorsal fin

spinal cord

spinal ganglion

notochord

dorsal aorta

somite

proctodeum

Figure 6.54

6/7-mm frog tadpole (stages 20-21), transverse section through the proctodeum (100X).

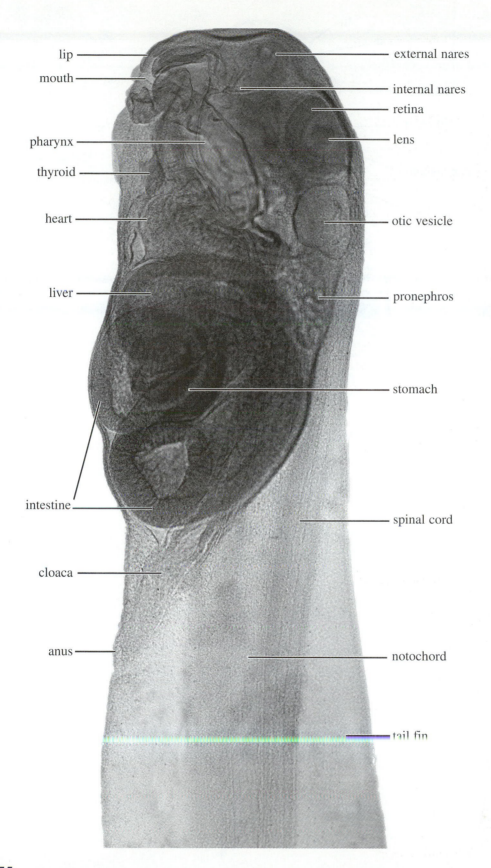

lip — external nares
mouth — internal nares
retina
pharynx — lens
thyroid
heart — otic vesicle
liver — pronephros
stomach
intestine — spinal cord
cloaca
anus — notochord
tail fin

Figure 6.55

10-mm frog tadpole (stage 24), whole mount (55X).

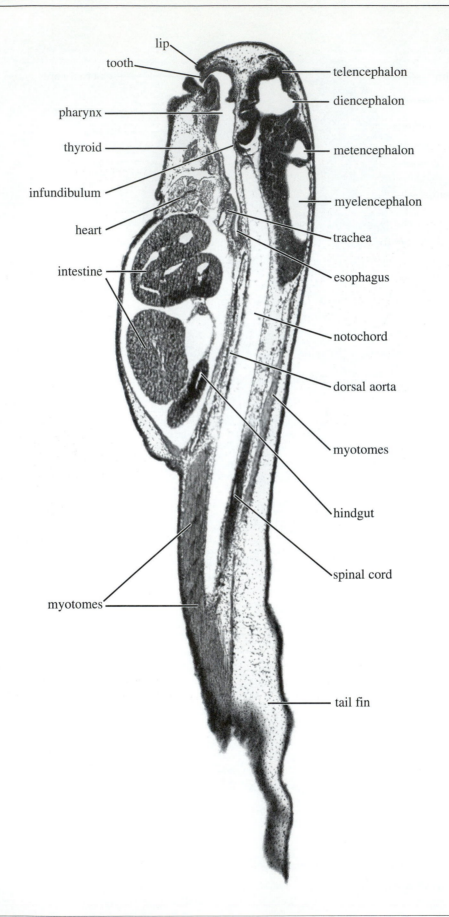

Figure 6.56

10-mm frog tadpole (stage 24), sagittal section (35X).

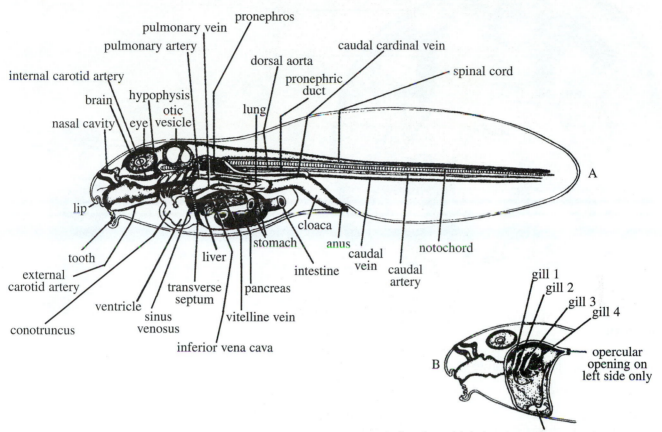

Figure 6.57

Drawing of the 10/18-mm frog tadpole. A, whole tadpole from left side; B, branchial chamber on left side. Aortic arches 3-6 function as part of the branchial mechanism; aortic arches 1 and 2 are vestigial in the frog. Arrows in B show water currents.

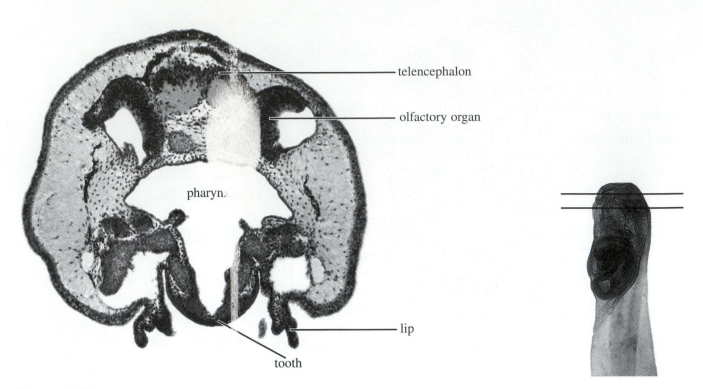

Figure 6.58

10-mm frog tadpole (stage 24), transverse section through the olfactory organ (50X).

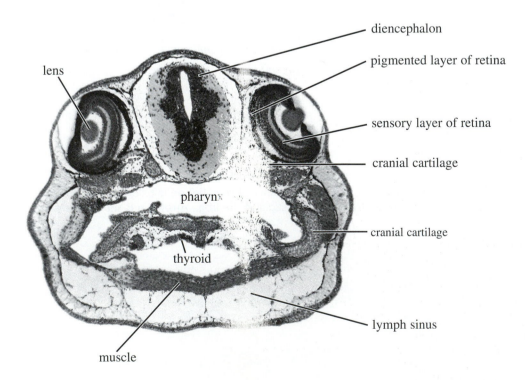

Figure 6.59

10-mm frog tadpole (stage 24), transverse section through the eyes (50X).

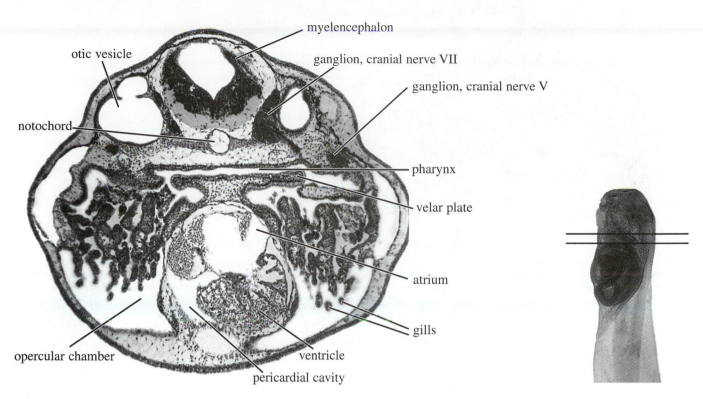

Figure 6.60

10-mm frog tadpole (stage 24), transverse section through the heart (50X).

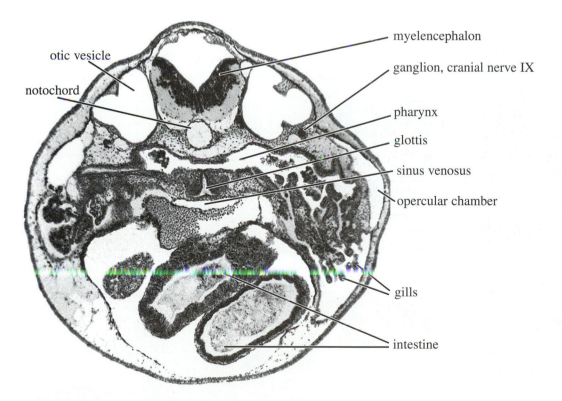

Figure 6.61

10-mm frog tadpole (stage 24), transverse section through the glottis (50X).

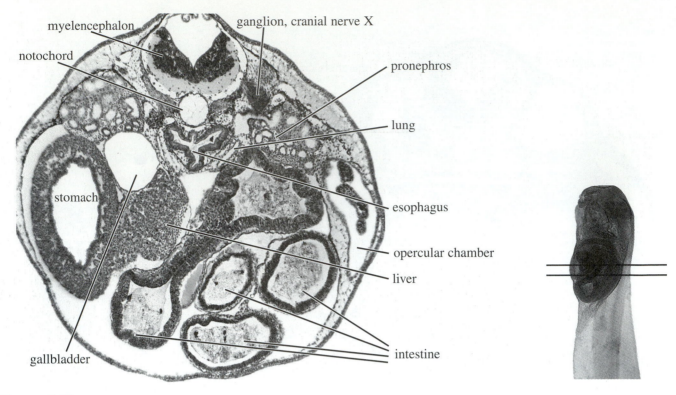

Figure 6.62

10-mm frog tadpole (stage 24), transverse section through the pronephros (50X).

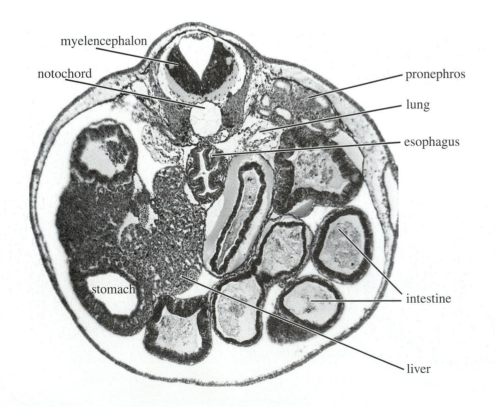

Figure 6.63

10-mm frog tadpole (stage 24), transverse section through the liver (50X).

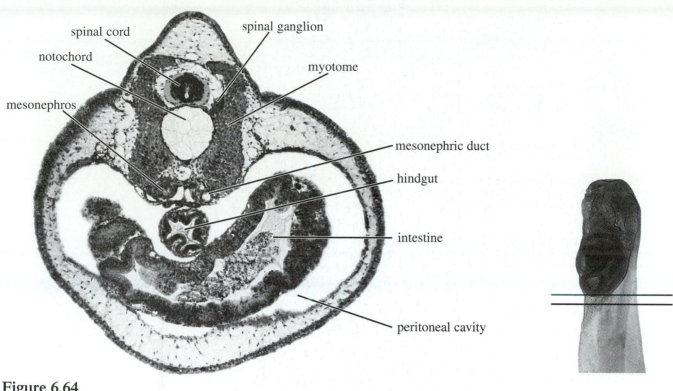

spinal cord

spinal ganglion

notochord

myotome

mesonephros

mesonephric duct

hindgut

intestine

peritoneal cavity

Figure 6.64

10-mm frog tadpole (stage 24), transverse section through the mesonephros (50X).

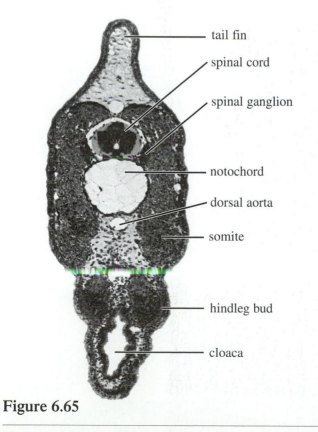

tail fin

spinal cord

spinal ganglion

notochord

dorsal aorta

somite

hindleg bud

cloaca

Figure 6.65

10-mm frog tadpole (stage 24), transverse section through the cloaca (50X).

Chapter 7

Avian Development

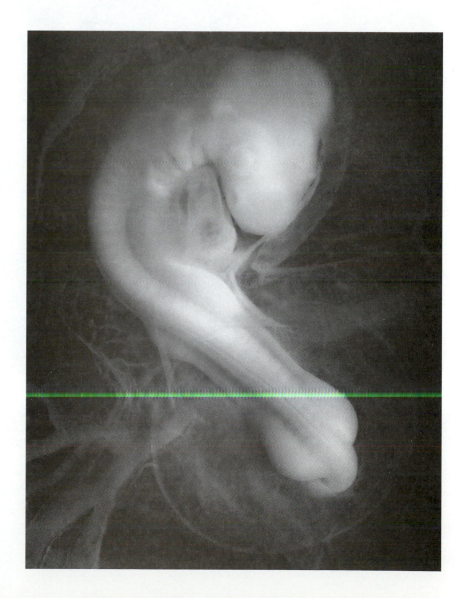

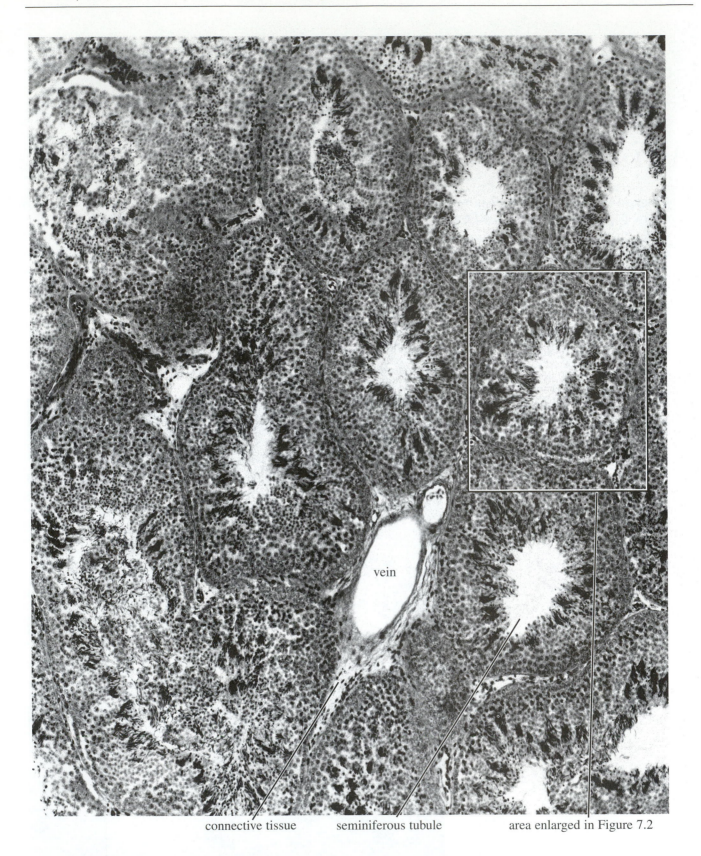

vein

connective tissue   seminiferous tubule   area enlarged in Figure 7.2

Figure 7.1

Chicken testis, section (280X).

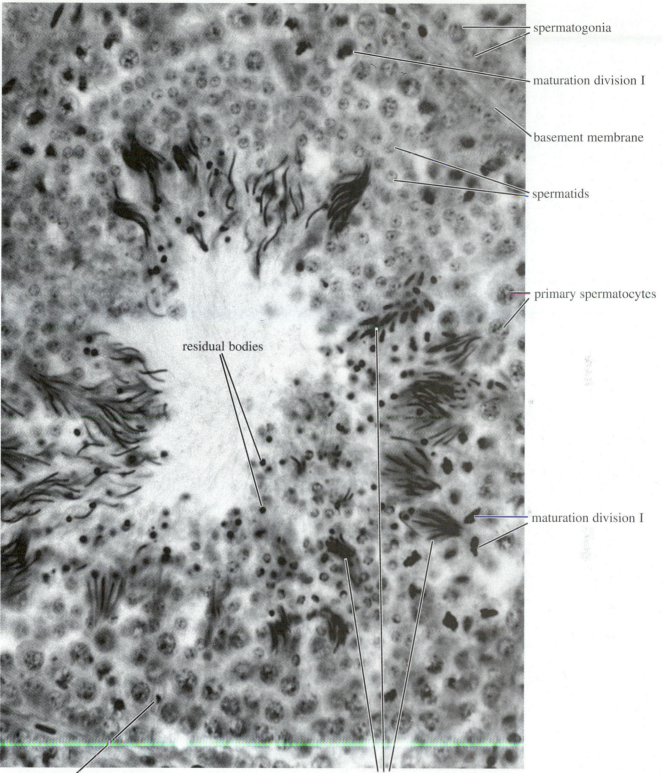

spermatogonia

maturation division I

basement membrane

spermatids

primary spermatocytes

residual bodies

maturation division I

Sertoli cell

immature sperm

Figure 7.2

Chicken testis, section, area indicated by the rectangle in Figure 7.1 (865X).

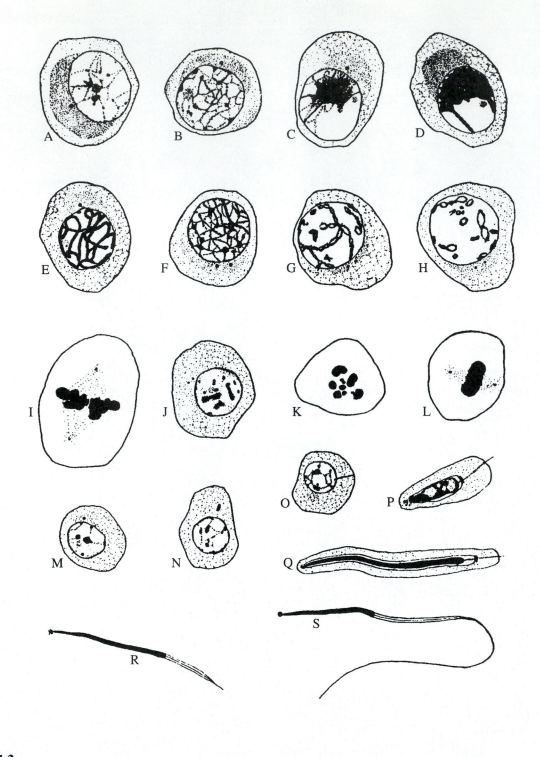

Figure 7.3

Spermatogenesis in the chicken. A-I, primary spermatocytes; A, leptotene; C, zygotene; D, E, pachytene; F, diffuse stage; G, H, diplotene; I, metaphase of first maturation division; J-L, secondary spermatocytes; J, interphase of secondary spermatocyte; K, L, metaphase of second maturation division; M, spermatid; N-S, spermiogenesis.

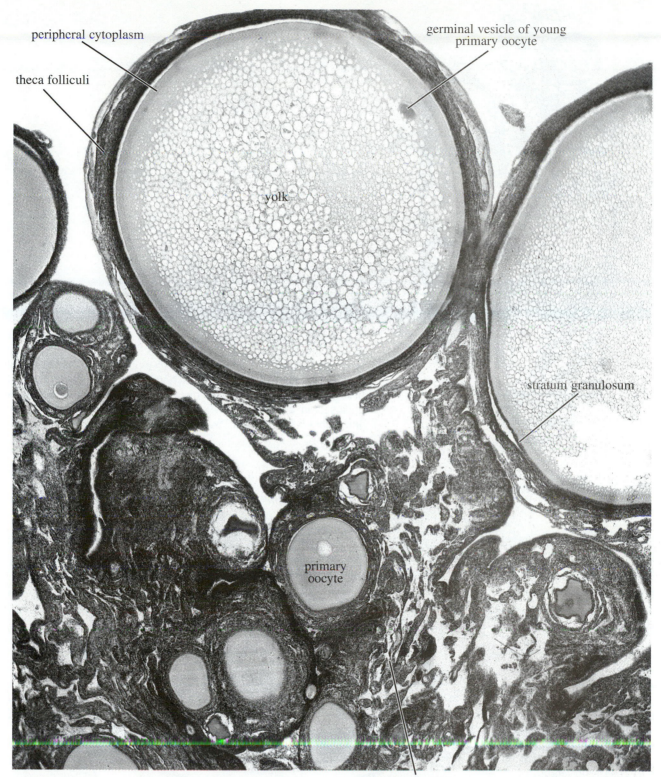

peripheral cytoplasm

theca folliculi

germinal vesicle of young
primary oocyte

yolk

stratum granulosum

primary
oocyte

connective tissue

Figure 7.4

Chicken ovary, section (50X).

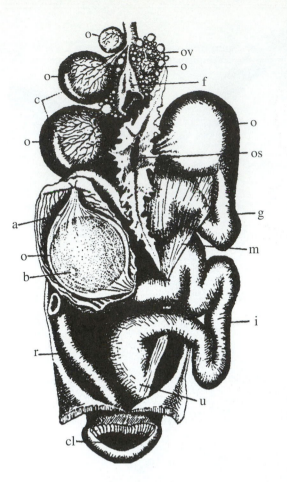

Figure 7.5

The female reproductive system of the fowl. The figure shows two eggs in the oviduct, though normally only one is in the oviduct at a time. a, albumen (dense layer); b, blastoderm; c, cicatrix; cl, cloaca; f, follicle from which the egg has been discharged; g, glandular portion of the oviduct; i, isthmus; o, ovarian ova in different stages of growth, each enclosed in a follicle richly supplied with blood vessels; os, ostium (infundibulum) of oviduct; ov, ovary; r, rectum; u, uterus.

Figure 7.6

Growth and composition of the hen's egg before incubation. The cells of the blastoderm are drawn relatively too large in size and too few in number. White yolk is lightly stippled, yellow yolk is heavily stippled (the central white yolk was called the latebra of Purkinje, and that below the blastoderm, the nucleus of Pander). Egg yolk, being relatively lighter than egg white, tends to float toward the highest part of the shell.

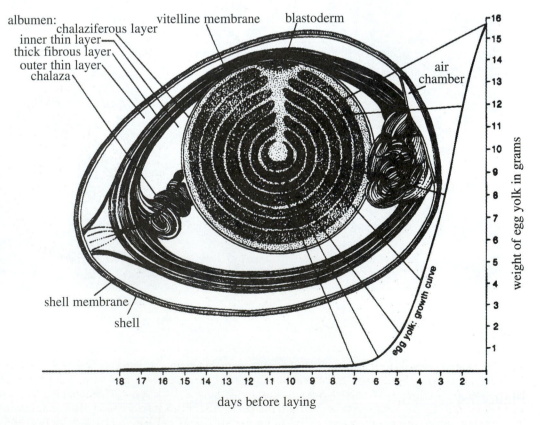

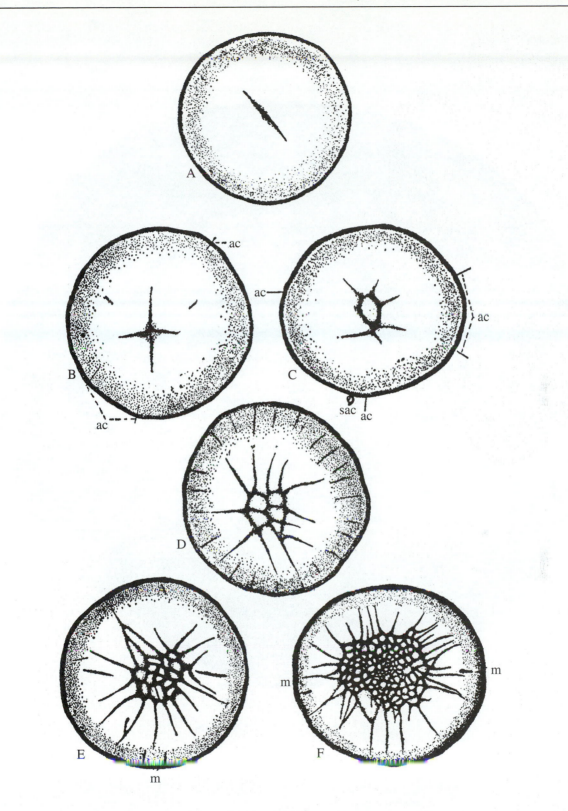

Figure 7.7

Surface views of the blastoderm of the hen's egg showing cleavage. A, surface view of the first cleavage furrow (3 hours after fertilization); the dark border represents the inner margin of the periblast; B, 4 cells, 3.25 hours after fertilization; C, 8 cells, 4 hours after fertilization; D, 17 cells, 4-5 hours after fertilization; E, 34 cells, 4.75 hours after fertilization; F, 154 cells (in surface view) of which 123 are central cells and 31 are marginal; 7 hours after fertilization; the blastoderm at this time averages 3 cells in thickness. ac, accessory cleavage; m, radial furrow; sac, small cell formed by accessory cleavage furrows.

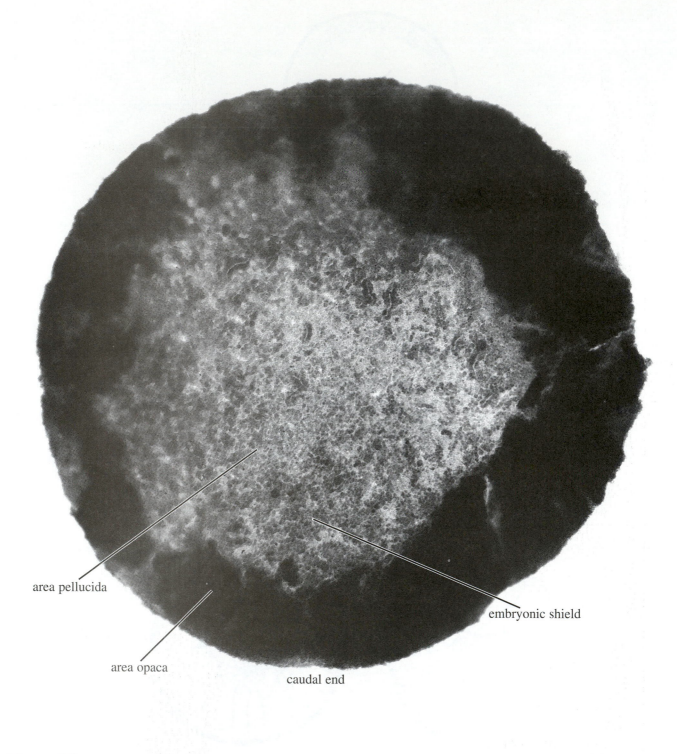

area pellucida

area opaca

caudal end

embryonic shield

Figure 7.8

Chick embryo, stage 1, unincubated blastoderm, whole mount (50X).

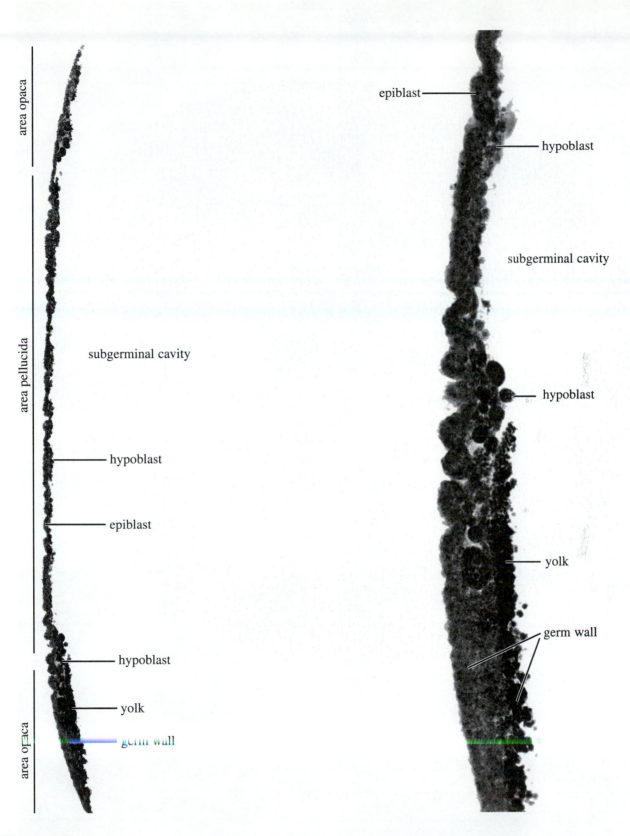

area opaca

area pellucida

subgerminal cavity

hypoblast

epiblast

hypoblast

yolk

area opaca

 germ wall

epiblast

hypoblast

subgerminal cavity

hypoblast

yolk

germ wall

Figure 7.9

Chick embryo, stage 1, unincubated blastoderm, sagittal section (60X).

Figure 7.10

Enlargement of caudal end of Figure 149 (150X).

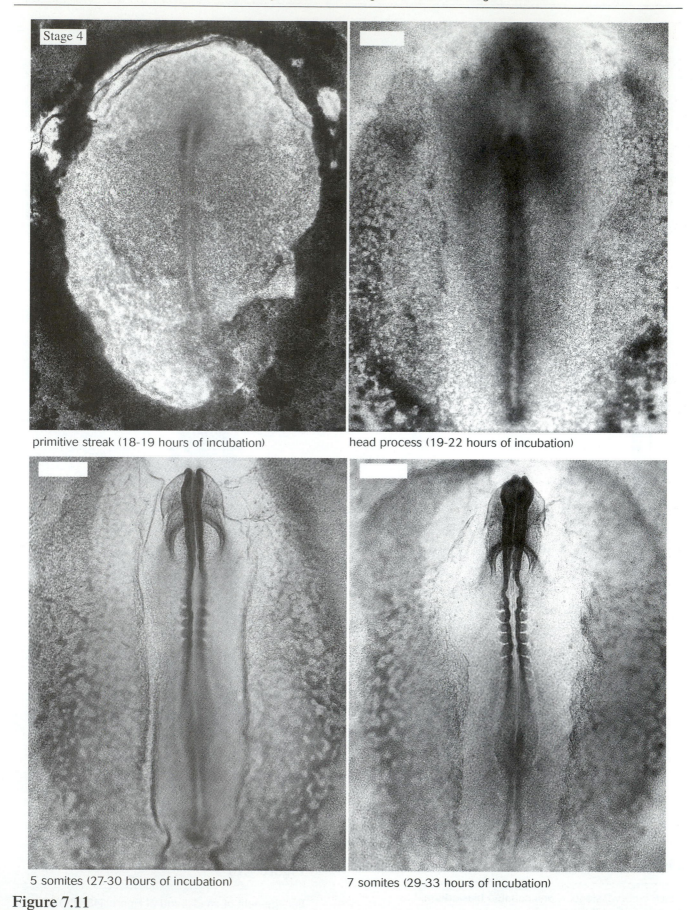

Stage 4

primitive streak (18-19 hours of incubation)

head process (19-22 hours of incubation)

5 somites (27-30 hours of incubation)

7 somites (29-33 hours of incubation)

Figure 7.11

Chick embryos, whole mounts (30X).

Table 2

Normal Stages of Chick Development

Because chick embryos develop at different rates under the same environmental conditions, designating developmental stages entirely by incubation times is unsatisfactory. A series of standard or "normal" stages has been established based on developmental features. Stage descriptions are adapted from Hamburger, V. and Hamilton, H.L., A series of normal stages in the development of the chick embryo. *Journal of Morphology* 88:49 (1951).

Stage 1. **Prestreak** (0-5 hours of incubation): The embryonic shield may be visible, but the primitive streak has not formed. **Figure 7.8**

Stage 2. **Initial Streak** (6-7 hours of incubation): The primitive streak appears as a short conical thickening at the border of the area pellucida.

Stage 3. **Intermediate Streak** (12-13 hours): The primitive streak extends to the center of the area pellucida.

Stage 4. **Definitive Streak** (18-19 hours): The primitive streak is at maximum length, with a primitive groove, primitive pit, and Hensen's node present. The area pellucida is shaped like a pear, and the primitive streak extends two-thirds to three-fourths of its length. **Figure 7.11.**

Stage 5. **Head Process** (19-22 hours): The notochord is visible, but the head fold has not formed. **Figures 7.11, 7.14.**

Stage 6. **Head Fold** (23-25 hours): The head fold is visible, but somites have not formed.

Stage 7. **One Somite** (23-26 hours): One pair of somites is visible, and neural folds are present.

Stage 8. **Four Somites** (26-29 hours): The neural folds are beginning to fuse, and blood islands are present. **Figures 7.11, 7.19, 7.20.**

Stage 9. **Seven Somites** (29-33 hours): The optic vesicles are present, and the heart tubes are beginning to fuse. **Figure 7.11.**

Stage 10. **Ten Somites** (33-38 hours): Three primary brain vesicles are visible. The optic vesicles are not constricted.

Stage 11. **Thirteen somites** (40-45 hours): Five neuromeres are visible in the hindbrain. The optic vesicles are constricted at their base, and the heart is bent to the right. **Figure 7.30.**

Stage 12. **Sixteen Somites** (45-49 hours): The telencephalon is visible; the auditory pits are deep; the heart is S-shaped. The head fold of the amnion covers the telencephalon.

Stage 13. **Nineteen Somites** (48-52 hours): The head is turning to the right; the telencephalon is enlarged; the amnionfold covers the head to the hindbrain. Cranial and cervical flexures are present.

Stage 14. **Twenty-Two Somites** (50-53 hours): The cranial flexure equals about 90°; branchial arches 1 and 2 and grooves 1 and 2 are distinct; the optic vesicles are invaginated as optic cups, and the lens placodes are present.

Stage 15. (50-55 hours): The cranial flexure is more than 90°; branchial arch 3 and groove 3 are distinct; the optic cups are well formed. **Figure 7.43.**

Stage 16. (51-56 hours): The wing buds are visible; the tail buds is present; the leg buds are not yet visible.

Stage 17. (52-64 hours): The wing and leg buds are visible; the epiphysis is distinct; the nasal pits are forming; the allantois is not visible.

Stage 18. (3 days): The leg buds are slightly longer than the wing buds; the amnion is nearly or completely closed; the cervical flexure equals about 90°; the maxillary processes and 4th grooves are indistinct or absent; the allantois is visible. **Figure 7.63.**

Stage 19. (3 to 3-1/2 days): The maxillary processes are distinct and are as long as the mandibular processes; the allantois is a small pocket but not yet vesicular; the eyes are unpigmented. **Figure 7.65.**

Stage 20. (3 to 3-1/2 days): The trunk is straight; the allantois is vesicular and about as large as the midbrain; the eye is slightly pigmented. **Figure 7.84.**

Stage 21. (3-1/2 days): The limb buds are slightly asymmetrical; their axis is directed caudally. The maxillary processes are longer than the mandible, extending to the middle of the eye. The 4th arch and groove are distinct; the allantois extends to the head; eye pigmentation is distinct.

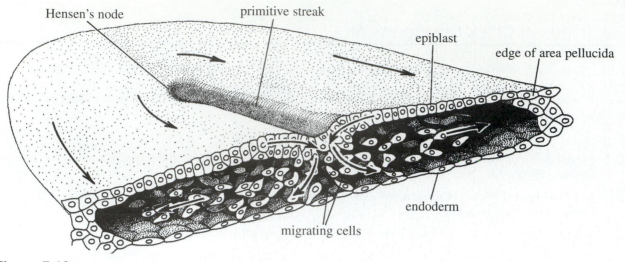

Figure 7.12

Rostral half of the area pellucida of a chick embryo cut transversely to show the migration of mesodermal and endodermal cells from the primitive streak.

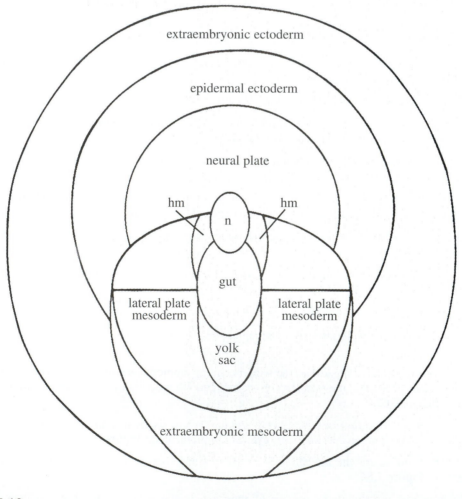

Figure 7.13

Fate map of the chick embryo at the beginning of gastrulation. n, notochord; hm, head mesenchyme.

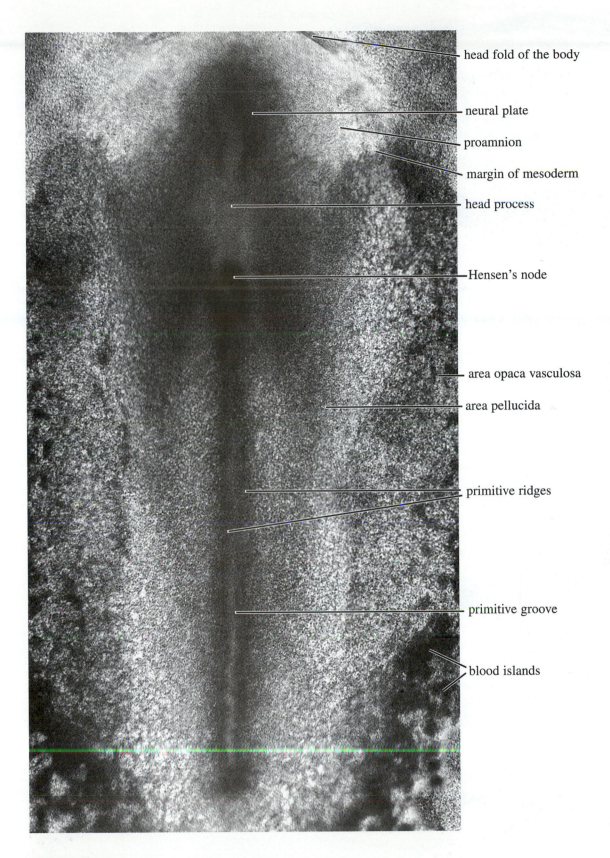

head fold of the body

neural plate

proamnion

margin of mesoderm

head process

Hensen's node

area opaca vasculosa

area pellucida

primitive ridges

primitive groove

blood islands

Figure 7.14

Chick embryo, head process stage (stage 5), whole mount (50X).

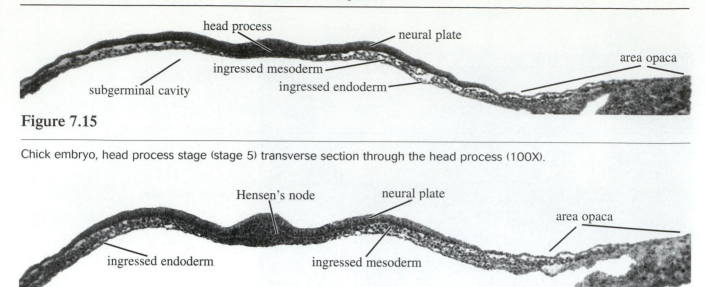

Figure 7.15

Chick embryo, head process stage (stage 5) transverse section through the head process (100X).

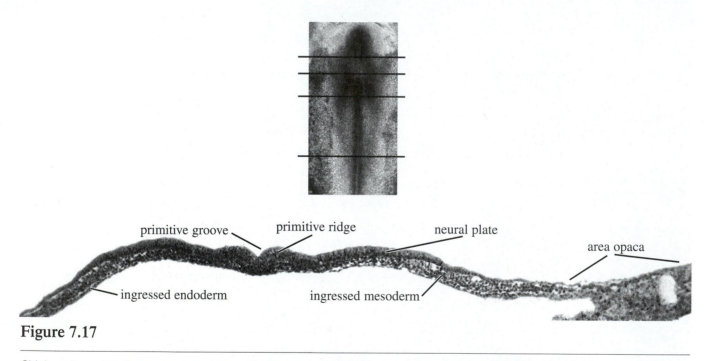

Figure 7.16

Chick embryo, head process stage (stage 5) transverse section through Hensen's node (100X).

Figure 7.17

Chick embryo, head process stage (stage 5) transverse section through the rostral primitive streak (100X).

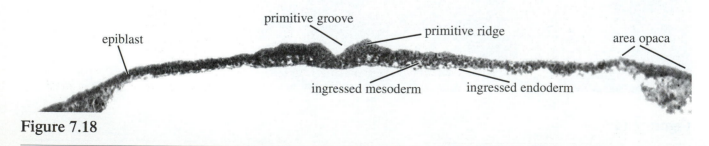

Figure 7.18

Chick embryo, head process stage (stage 5) transverse section through the caudal primitive streak (100X).

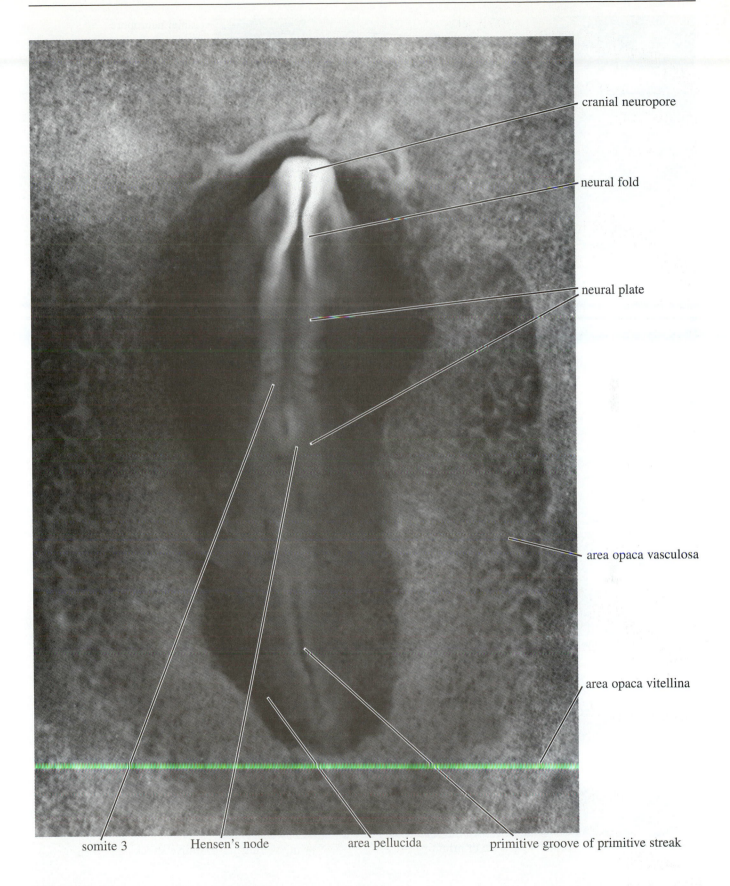

cranial neuropore

neural fold

neural plate

area opaca vasculosa

area opaca vitellina

somite 3   Hensen's node   area pellucida   primitive groove of primitive streak

Figure 7.19

4-somite chick embryo (stage 8), opaque whole mount, incident illumination (50X).

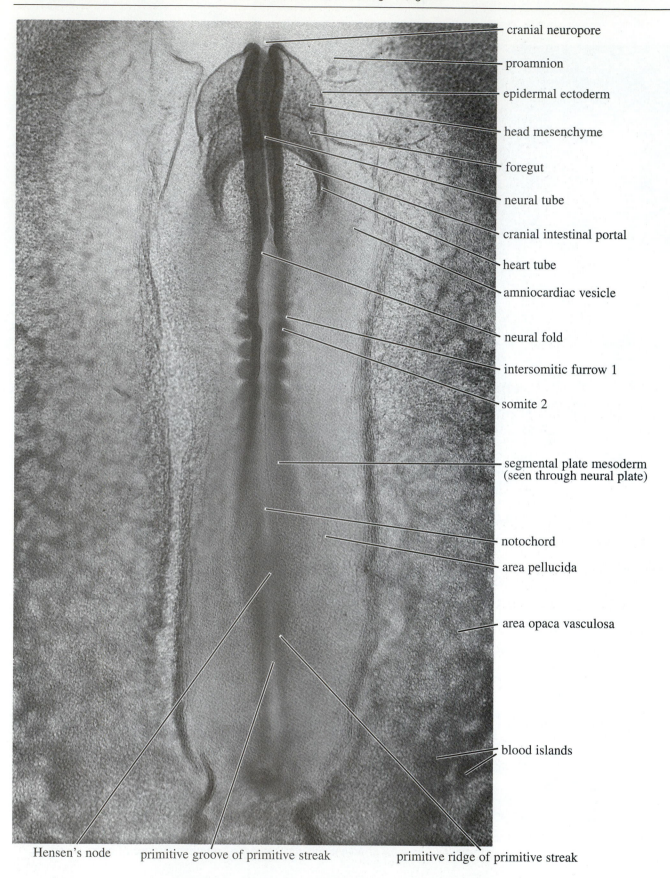

cranial neuropore

proamnion

epidermal ectoderm

head mesenchyme

foregut

neural tube

cranial intestinal portal

heart tube

amniocardiac vesicle

neural fold

intersomitic furrow 1

somite 2

segmental plate mesoderm
(seen through neural plate)

notochord

area pellucida

area opaca vasculosa

blood islands

Hensen's node   primitive groove of primitive streak   primitive ridge of primitive streak

Figure 7.20

5-somite chick embryo (stage 8), whole mount (55X).

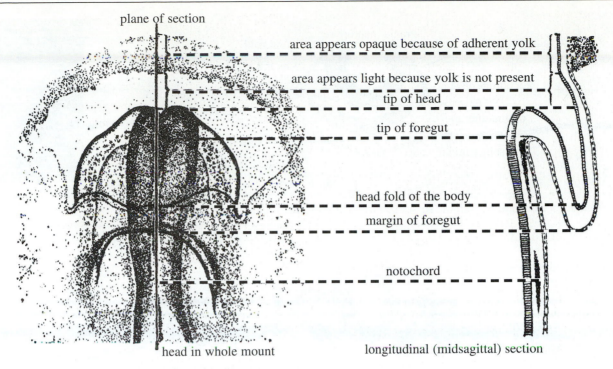

Figure 7.21

Relationship between the head of a 24-hour chick embryo viewed in whole mount and a sagittal section through the same region.

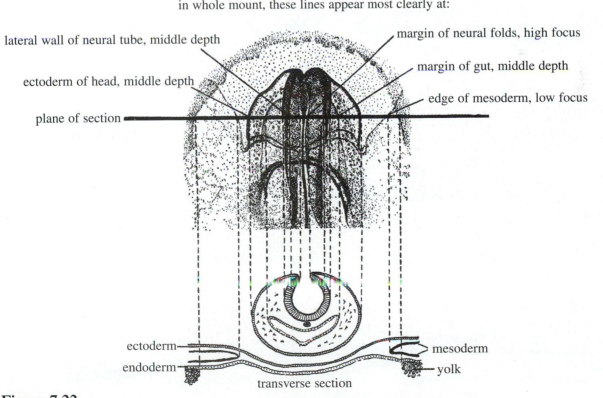

Figure 7.22

Relationship between the head of a 24-hour chick embryo viewed in whole mount and a transverse section through the same region.

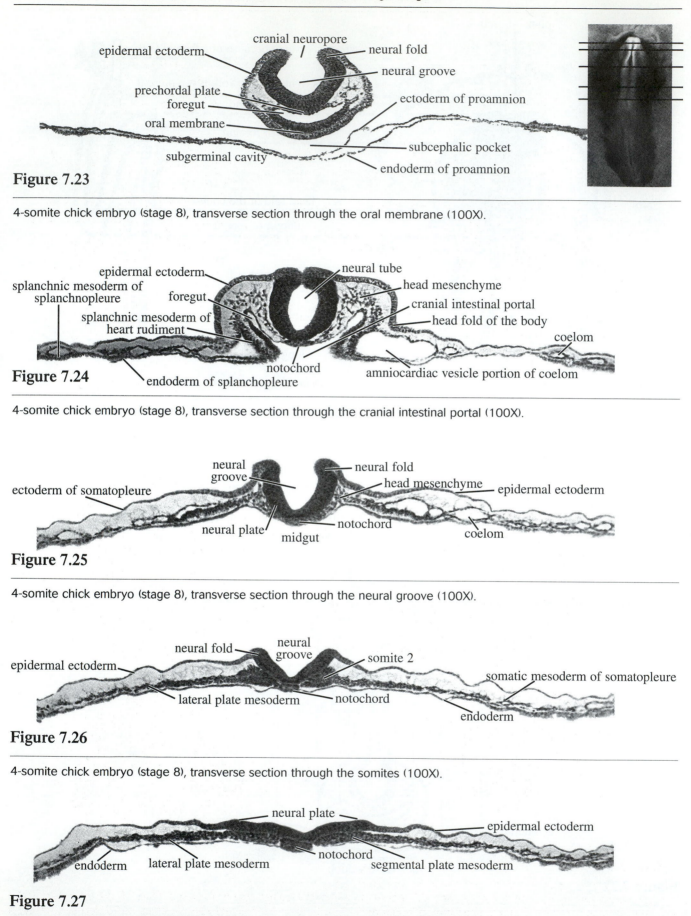

Figure 7.23

4-somite chick embryo (stage 8), transverse section through the oral membrane (100X).

Figure 7.24

4-somite chick embryo (stage 8), transverse section through the cranial intestinal portal (100X).

Figure 7.25

4-somite chick embryo (stage 8), transverse section through the neural groove (100X).

Figure 7.26

4-somite chick embryo (stage 8), transverse section through the somites (100X).

Figure 7.27

4-somite chick embryo (stage 8), transverse section through the neural plate (100X).

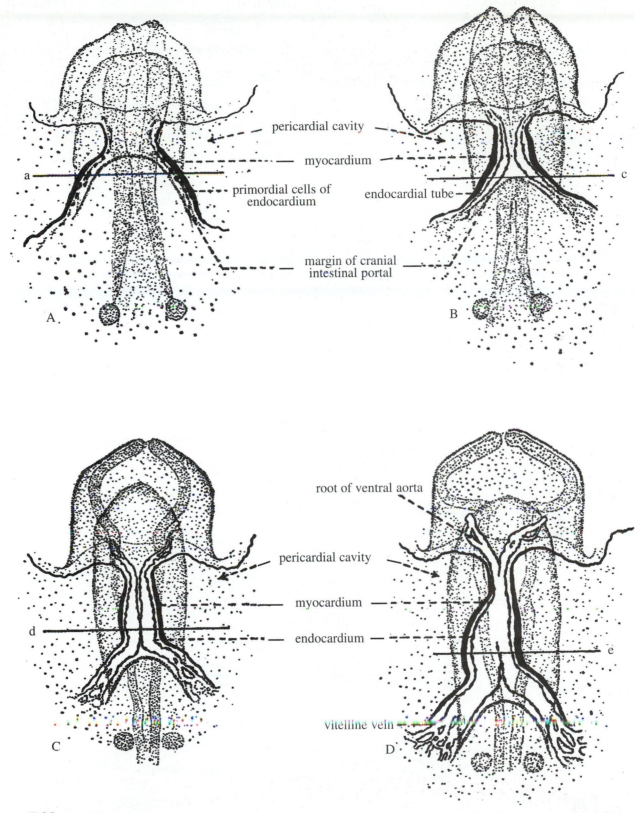

pericardial cavity

myocardium

a

primordial cells of
endocardium

endocardial tube

c

margin of cranial
intestinal portal

A.

B.

root of ventral aorta

pericardial cavity

myocardium

d

endocardium

e

vitelline vein

C.

D.

Figure 7.28

Ventral-view diagrams to show the origin and subsequent fusion of the paired primordia of the heart. The lines a, c, d and e indicate the planes of the sections diagrammed in Figure 7.29 A, C, D and E, respectively. A, chick of 25 hours; B, chick of 27 hours; C, chick of 28 hours; D, chick of 29 hours.

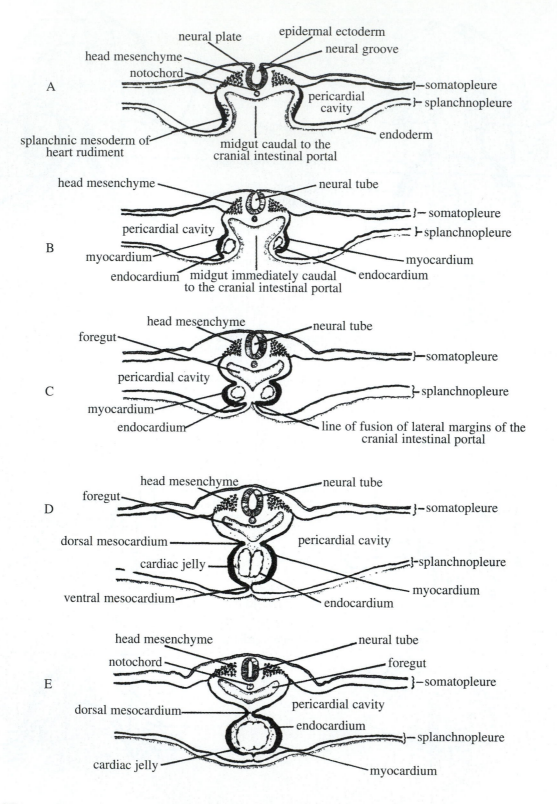

Figure 7.29

Diagrams of transverse sections through the heart region of chicks at various stages to show the formation of the heart. For the location of the sections see Figure 7.28. A, at 25 hours; B, at 26 hours; C, at 27 hours; D, at 28 hours; E, at 29 hours.

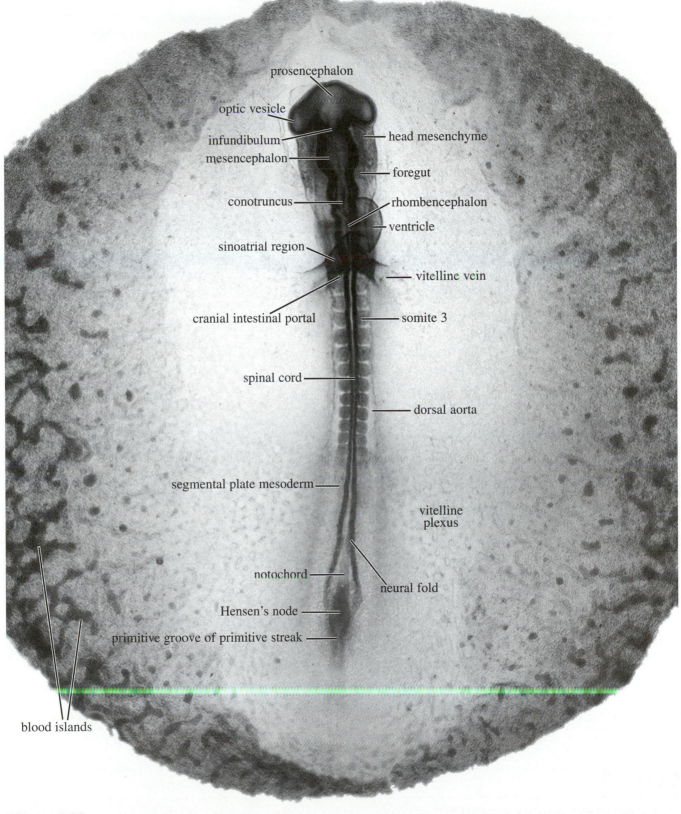

Figure 7.30

12-somite chick embryo (stage 11), whole mount (30X).

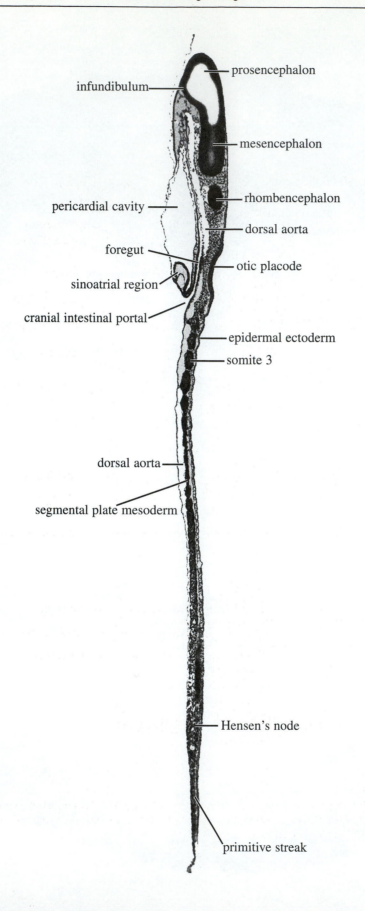

Figure 7.31

12-somite chick embryo (stage 11), sagittal section (30X).

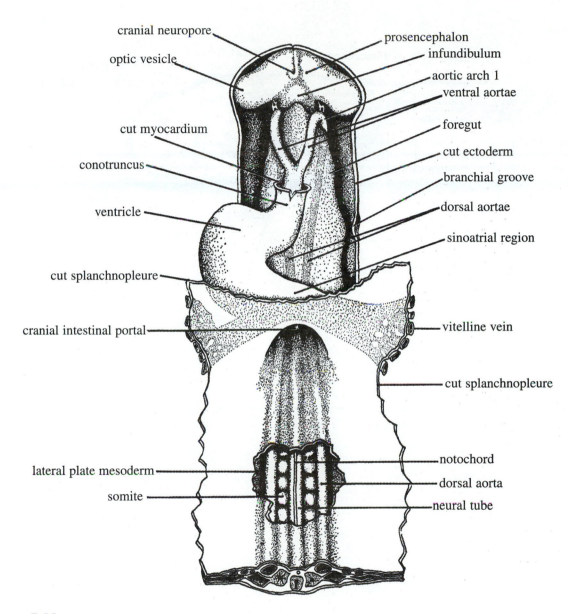

Figure 7.32

Diagrammatic ventral view of a dissection of a 13-somite chick embryo (stage 11). The splanchnopleure of the yolk sac rostral to the cranial intestinal portal, the ectoderm of the ventral surface of the head and the mesoderm of the pericardial region have been removed to show the underlying structures.

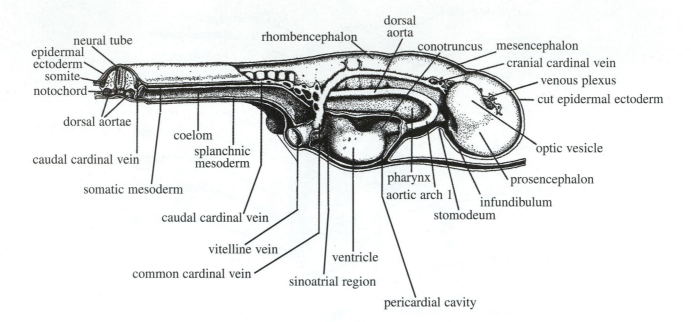

Figure 7.33

Diagrammatic lateral view of a dissection of a 12-somite chick (stage 11). The lateral body wall on the right side has been removed to show the internal structures. Note especially the relations of the pericardial region to that part of the coelom that lies farther caudally, and the small anastomosing channels of the developing caudal cardinal vein from which a single main vessel is later derived.

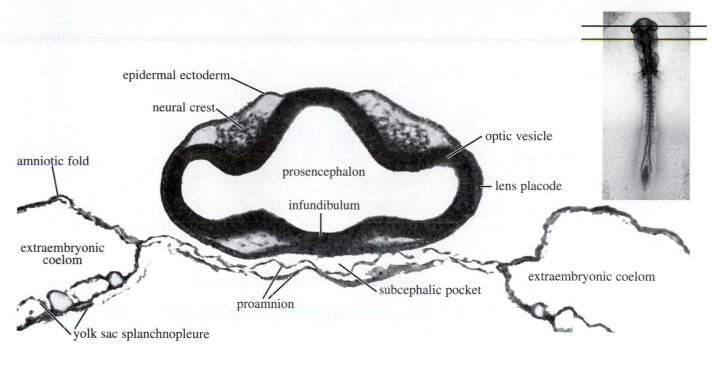

Figure 7.34

13-somite chick embryo (stage 11), transverse section through the optic vesicles (120X).

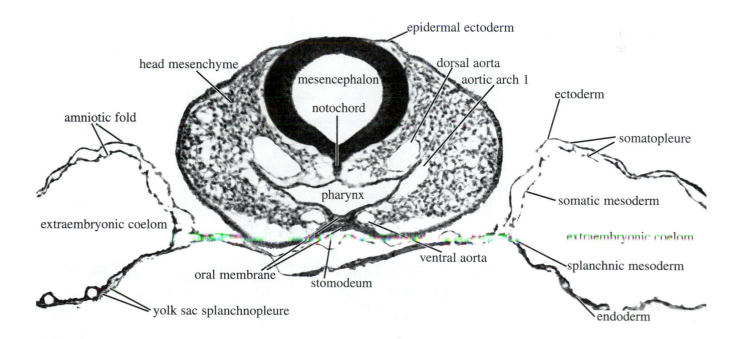

Figure 7.35

13-somite chick embryo (stage 11), transverse section through the oral membrane (120X).

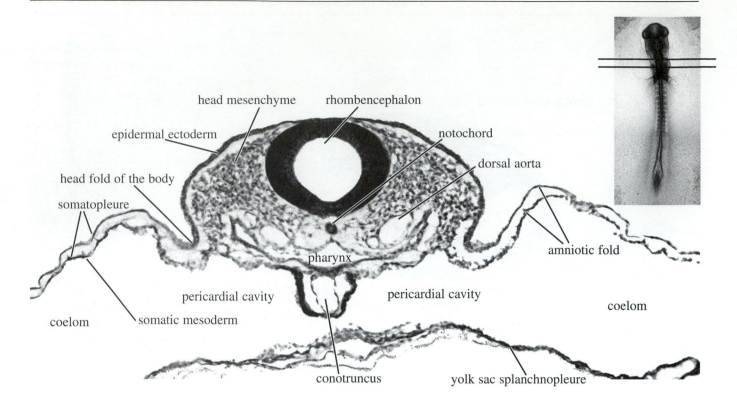

Figure 7.36

13-somite chick embryo (stage 11), transverse section through the conotruncus (120X).

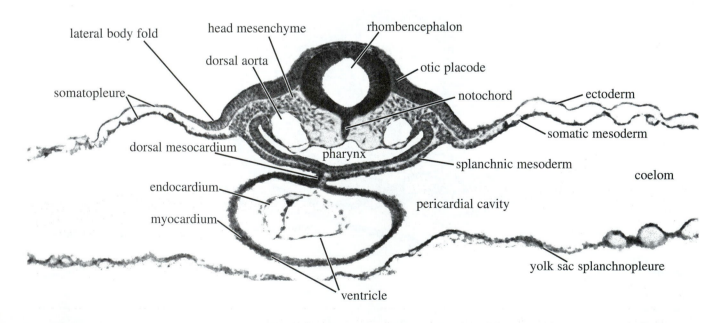

Figure 7.37

13-somite chick embryo (stage 11), transverse section through the ventricle (120X).

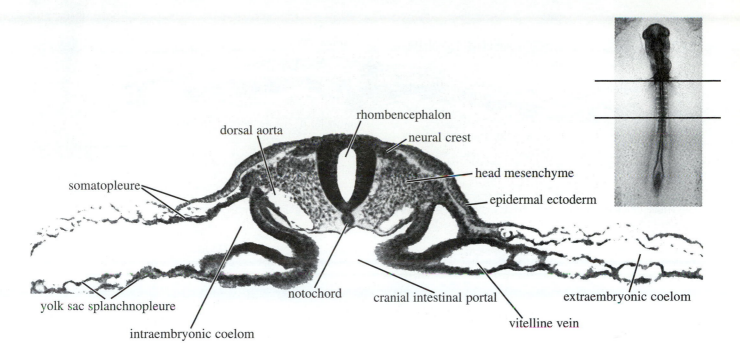

Figure 7.38

13-somite chick embryo (stage 11), transverse section through the cranial intestinal portal (120X).

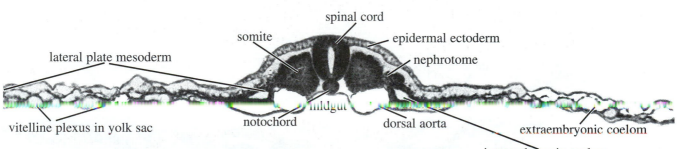

Figure 7.39

13-somite chick embryo (stage 11), transverse section through the midgut (120X).

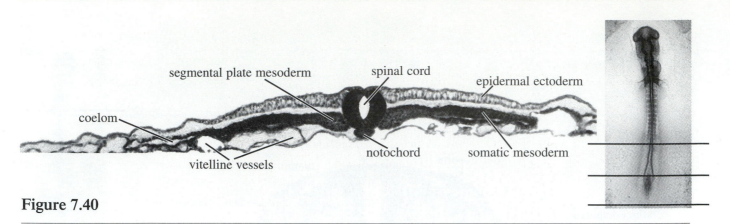

Figure 7.40

13-somite chick embryo (stage 11), transverse section through the segmental plate mesoderm (120X).

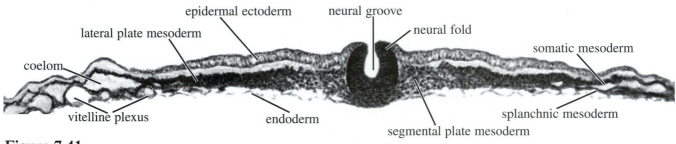

Figure 7.41

13-somite chick embryo (stage 11), transverse section through the neural groove (120X).

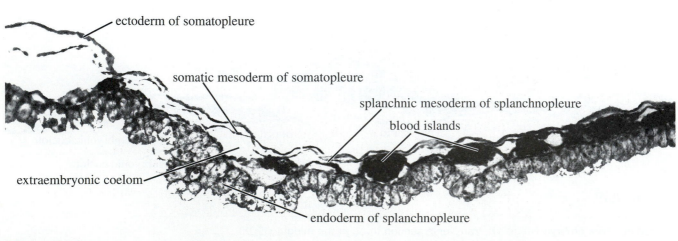

Figure 7.42

13-somite chick embryo (stage 11), transverse section through the area vasculosa of the area opaca (120X).

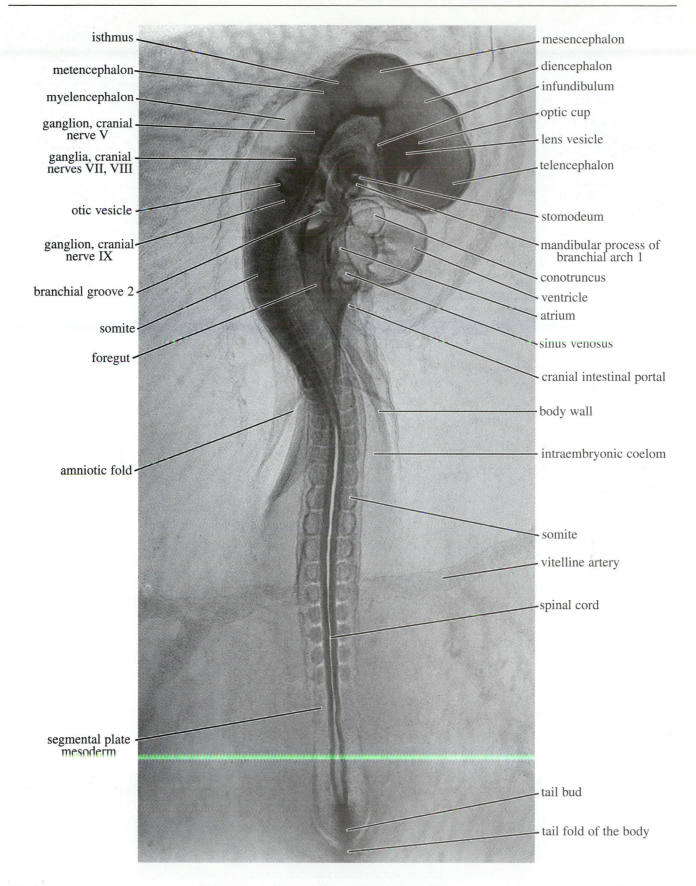

isthmus

metencephalon

myelencephalon

ganglion, cranial
nerve V

ganglia, cranial
nerves VII, VIII

otic vesicle

ganglion, cranial
nerve IX

branchial groove 2

somite

foregut

amniotic fold

segmental plate
mesoderm

mesencephalon

diencephalon

infundibulum

optic cup

lens vesicle

telencephalon

stomodeum

mandibular process of
branchial arch 1

conotruncus

ventricle

atrium

sinus venosus

cranial intestinal portal

body wall

intraembryonic coelom

somite

vitelline artery

spinal cord

tail bud

tail fold of the body

Figure 7.43

2-day chick embryo (stage 15), whole mount (25X).

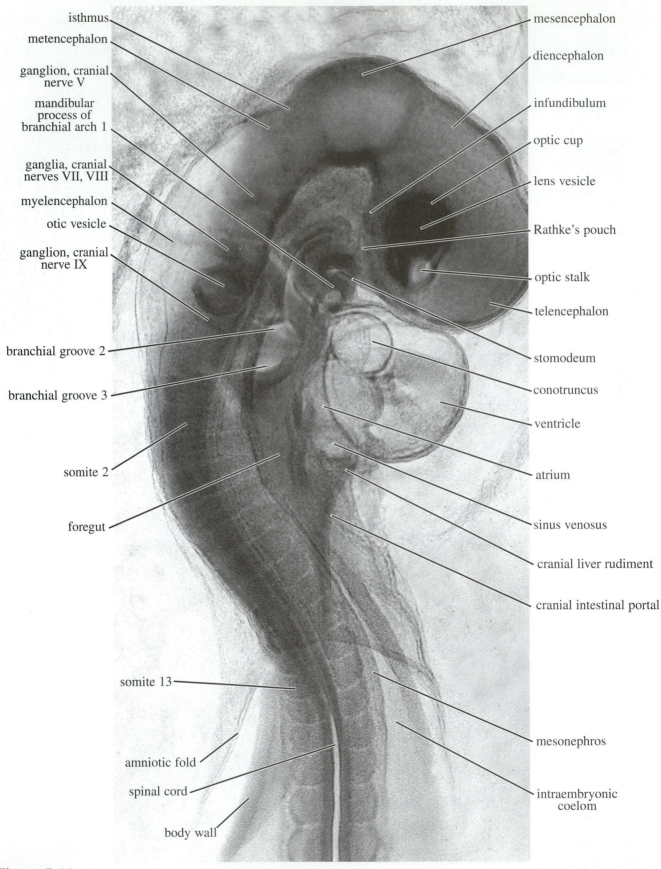

isthmus
metencephalon
ganglion, cranial nerve V
mandibular process of branchial arch 1
ganglia, cranial nerves VII, VIII
myelencephalon
otic vesicle
ganglion, cranial nerve IX
branchial groove 2
branchial groove 3
somite 2
foregut
somite 13
amniotic fold
spinal cord
body wall

mesencephalon
diencephalon
infundibulum
optic cup
lens vesicle
Rathke's pouch
optic stalk
telencephalon
stomodeum
conotruncus
ventricle
atrium
sinus venosus
cranial liver rudiment
cranial intestinal portal
mesonephros
intraembryonic coelom

Figure 7.44

Cranial half of 2-day chick embryo (stage 15), whole mount (50X).

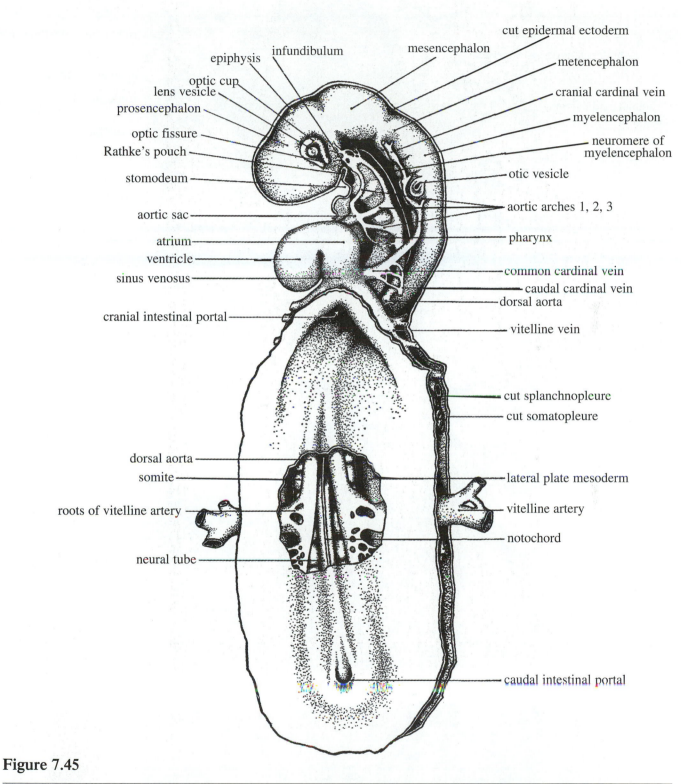

Figure 7.45

Diagram of a dissection of the chick embryo of about 50 hours. The splanchnopleure of the yolk sac cranial to the cranial intestinal portal, the ectoderm of the left side of the head and the mesoderm in the pericardial region have been dissected away. A window has been cut in the splanchnopleure of the dorsal wall of the midgut to show the origin of the vitelline artery.

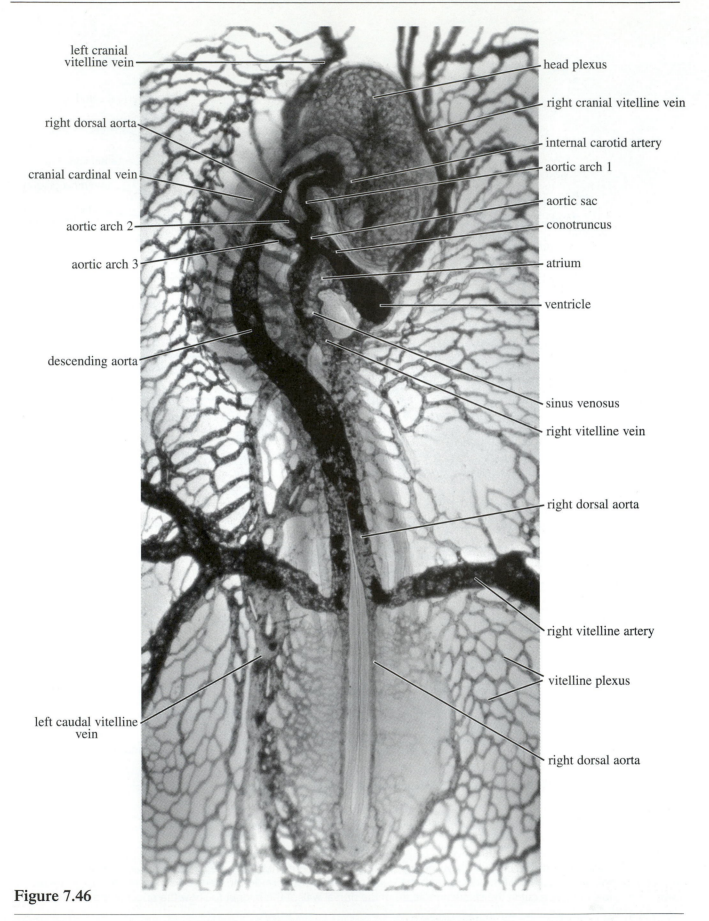

left cranial
vitelline vein

head plexus

right dorsal aorta

right cranial vitelline vein

cranial cardinal vein

internal carotid artery

aortic arch 1

aortic sac

aortic arch 2

conotruncus

aortic arch 3

atrium

ventricle

descending aorta

sinus venosus

right vitelline vein

right dorsal aorta

right vitelline artery

vitelline plexus

left caudal vitelline
vein

right dorsal aorta

Figure 7.46

2-day chick embryo (stage 15), whole mount, blood vessels injected with India ink (30X).

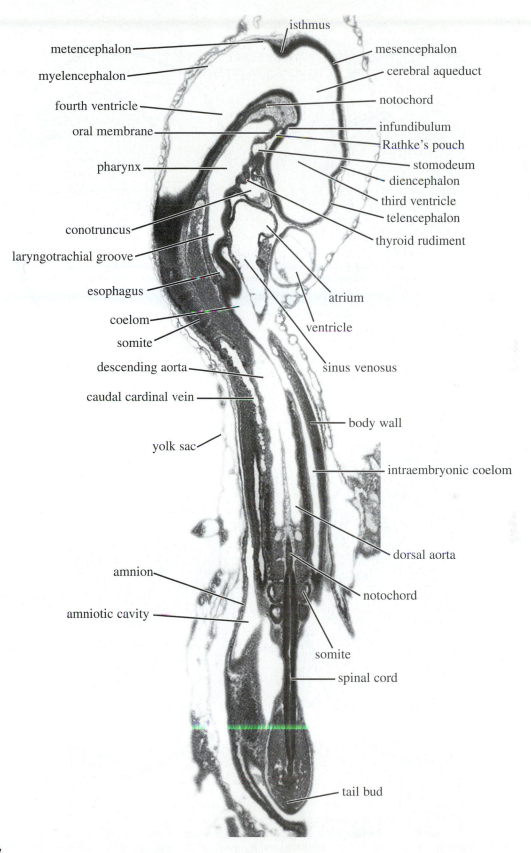

isthmus

metencephalon

myelencephalon

fourth ventricle

oral membrane

pharynx

conotruncus

laryngotrachial groove

esophagus

coelom

somite

descending aorta

caudal cardinal vein

yolk sac

amnion

amniotic cavity

mesencephalon

cerebral aqueduct

notochord

infundibulum

Rathke's pouch

stomodeum

diencephalon

third ventricle

telencephalon

thyroid rudiment

atrium

ventricle

sinus venosus

body wall

intraembryonic coelom

dorsal aorta

notochord

somite

spinal cord

tail bud

Figure 7.47

2-day chick embryo (stage 15), sagittal section (30X).

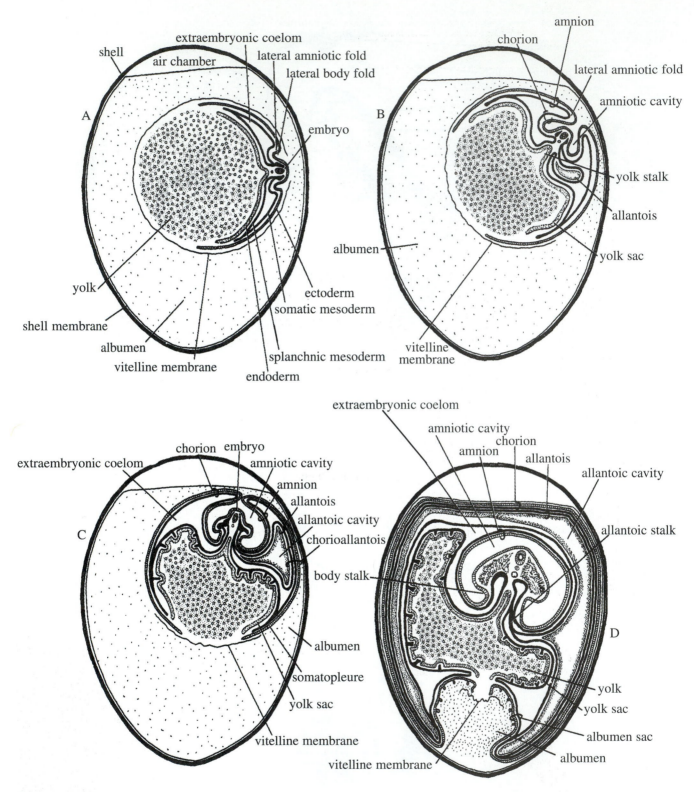

Figure 7.48

Diagrams showing the extraembryonic membranes of the chick. Each shows a section through the entire egg. The body of the embryo, being oriented approximately at right angles to the long axis of the egg, is cut transversely. The amnion and chorion are composed of somatopleure (ectoderm and somatic mesoderm) and the yolk sac and allantois are composed of splanchnopleure (endoderm and splanchnic mesoderm).   A, an embryo of 2 days of incubation; B, an embryo of 3 days of incubation; C, an embryo of 5 days of incubation; D, an embryo of 14 days of incubation.

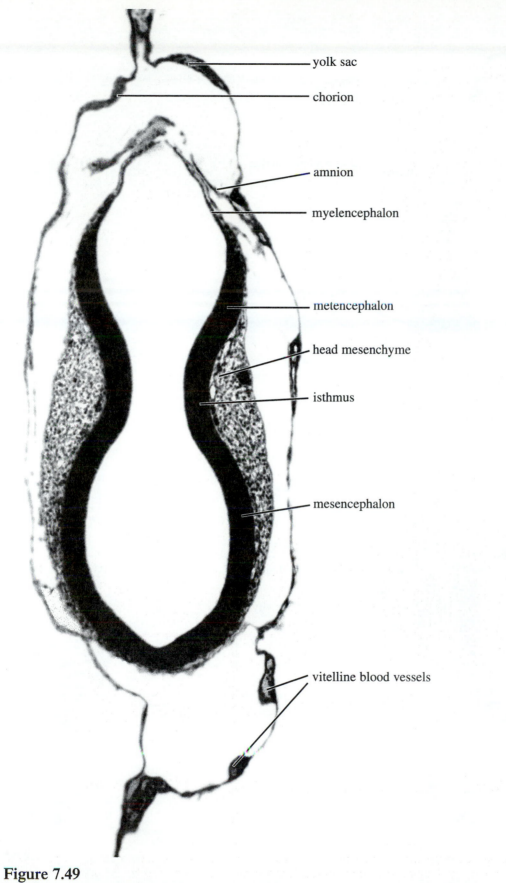

yolk sac

chorion

amnion

myelencephalon

metencephalon

head mesenchyme

isthmus

mesencephalon

vitelline blood vessels

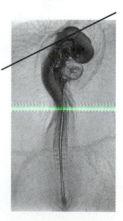

Figure 7.49

2-day chick embryo (stage 15), transverse section through the mesencephalon (110X).

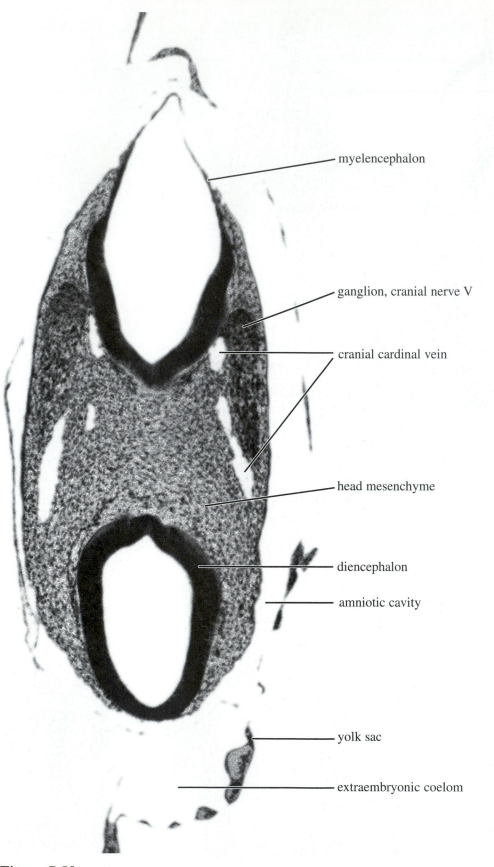

myelencephalon

ganglion, cranial nerve V

cranial cardinal vein

head mesenchyme

diencephalon

amniotic cavity

yolk sac

extraembryonic coelom

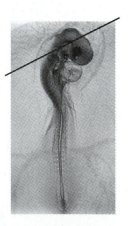

Figure 7.50

2-day chick embryo (stage 15), transverse section through the ganglion of cranial nerve V (110X).

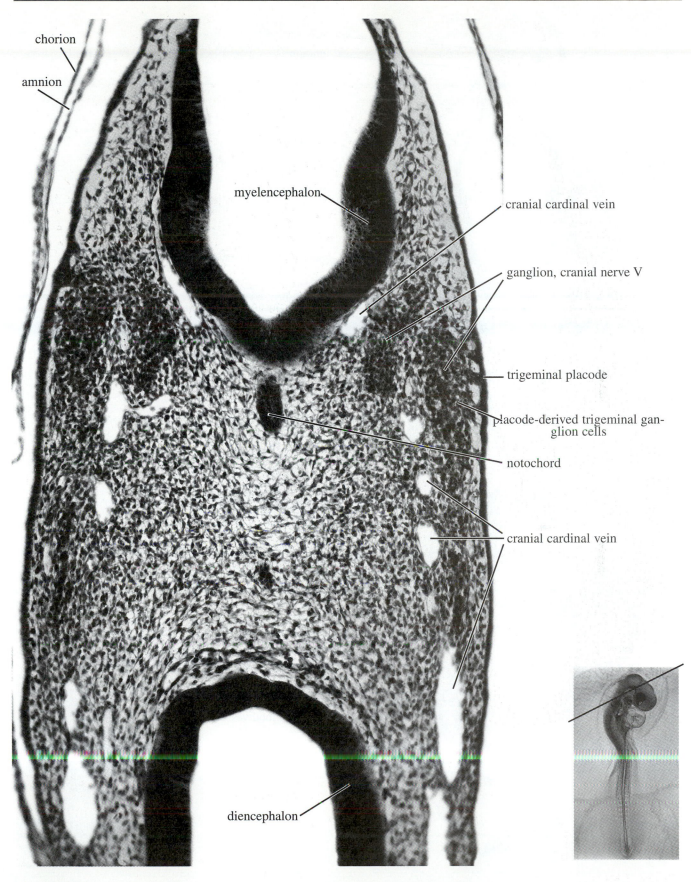

chorion

amnion

myelencephalon

cranial cardinal vein

ganglion, cranial nerve V

trigeminal placode

placode-derived trigeminal ganglion cells

notochord

cranial cardinal vein

diencephalon

Figure 7.51

2-day chick embryo (stage 15), transverse section through the trigeminal placode (180X).

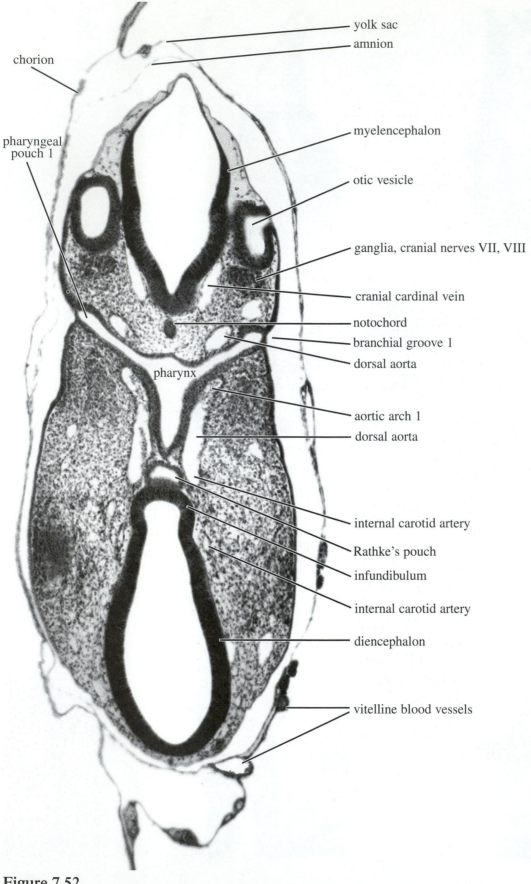

yolk sac

amnion

chorion

pharyngeal
pouch 1

myelencephalon

otic vesicle

ganglia, cranial nerves VII, VIII

cranial cardinal vein

notochord

branchial groove 1

dorsal aorta

pharynx

aortic arch 1

dorsal aorta

internal carotid artery

Rathke's pouch

infundibulum

internal carotid artery

diencephalon

vitelline blood vessels

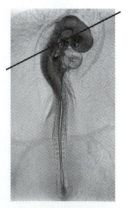

Figure 7.52

2-day chick embryo (stage 15), transverse section through the otic vesicle (110X).

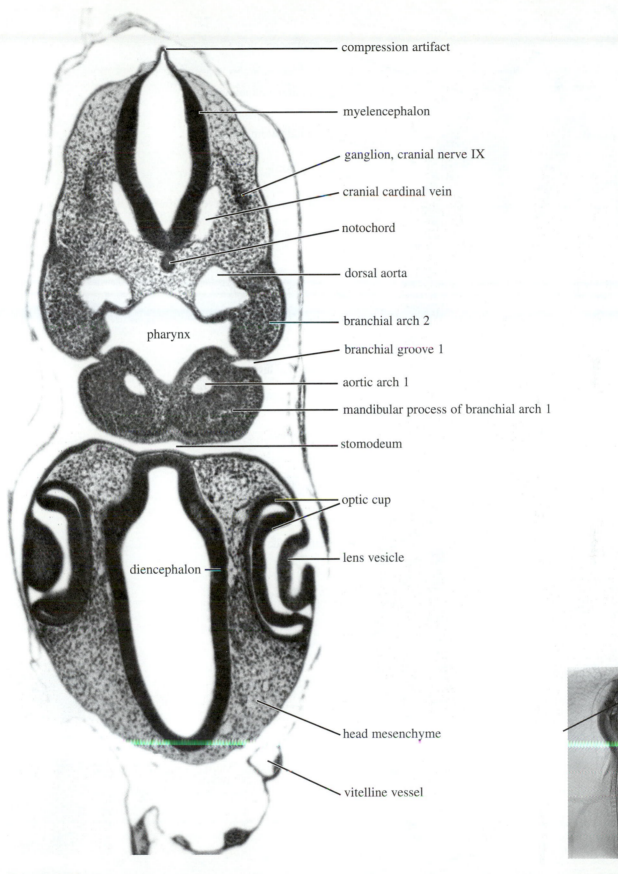

compression artifact

myelencephalon

ganglion, cranial nerve IX

cranial cardinal vein

notochord

dorsal aorta

branchial arch 2

branchial groove 1

aortic arch 1

mandibular process of branchial arch 1

stomodeum

optic cup

lens vesicle

pharynx

diencephalon

head mesenchyme

vitelline vessel

Figure 7.53

2-day chick embryo (stage 15), transverse section through the optic cups (110X).

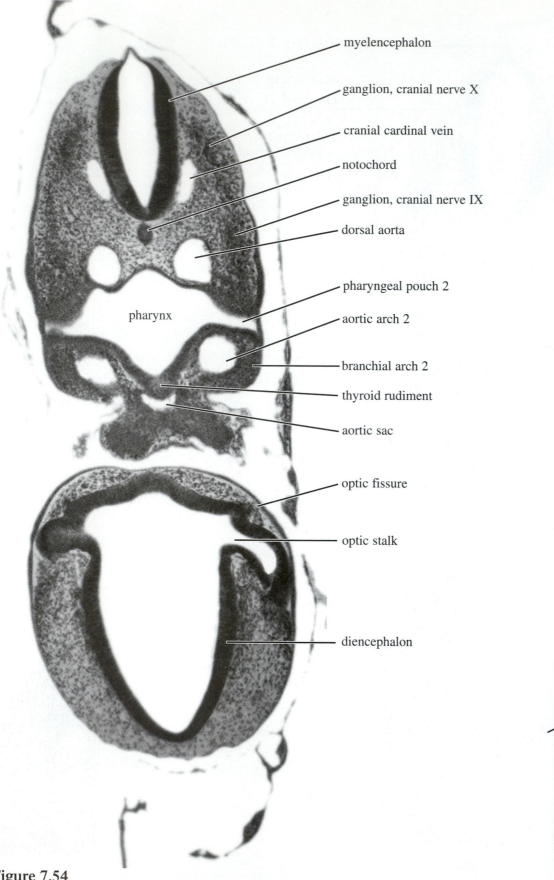

myelencephalon

ganglion, cranial nerve X

cranial cardinal vein

notochord

ganglion, cranial nerve IX

dorsal aorta

pharyngeal pouch 2

aortic arch 2

branchial arch 2

thyroid rudiment

aortic sac

optic fissure

optic stalk

diencephalon

pharynx

Figure 7.54

2-day chick embryo (stage 15), transverse section through the thyroid rudiment (110X).

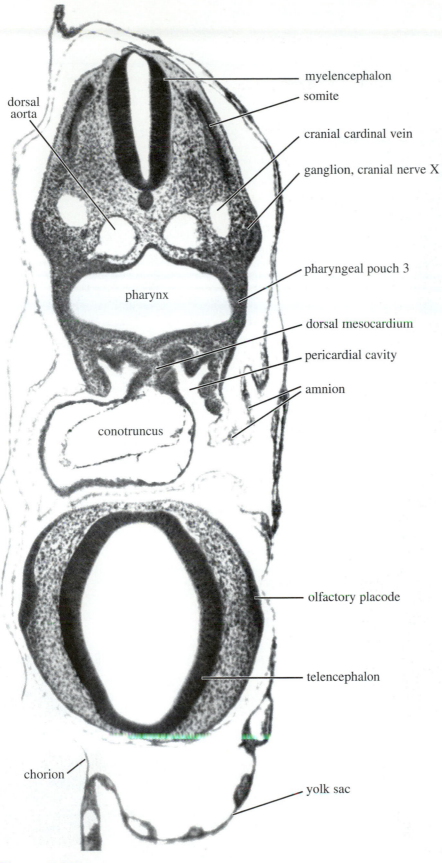

myelencephalon

somite

dorsal aorta

cranial cardinal vein

ganglion, cranial nerve X

pharyngeal pouch 3

pharynx

dorsal mesocardium

pericardial cavity

amnion

conotruncus

olfactory placode

telencephalon

chorion

yolk sac

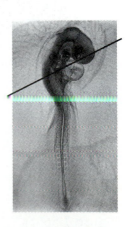

Figure 7.55

2-day chick embryo (stage 15), transverse section through the olfactory placodes (110X).

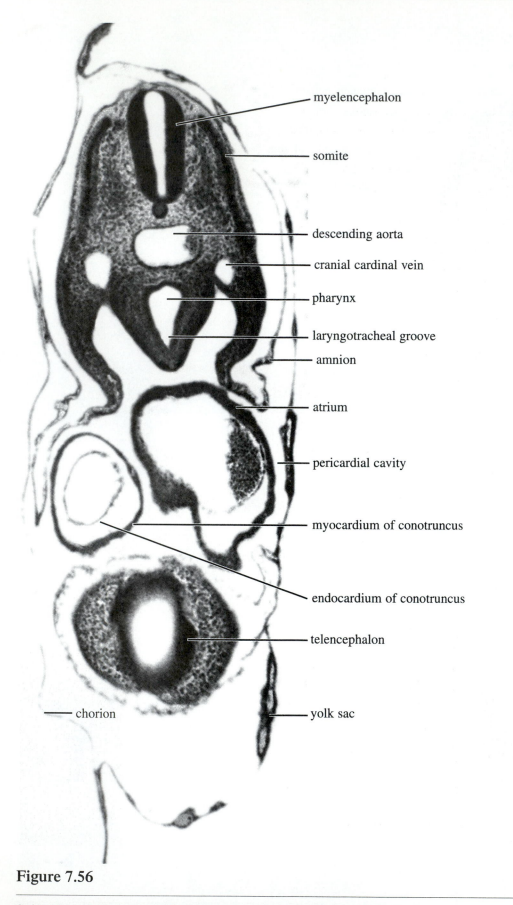

myelencephalon

somite

descending aorta

cranial cardinal vein

pharynx

laryngotracheal groove

amnion

atrium

pericardial cavity

myocardium of conotruncus

endocardium of conotruncus

telencephalon

chorion

yolk sac

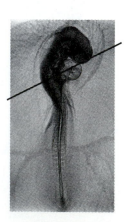

Figure 7.56

2-day chick embryo (stage 15), transverse section through the atrium (110X).

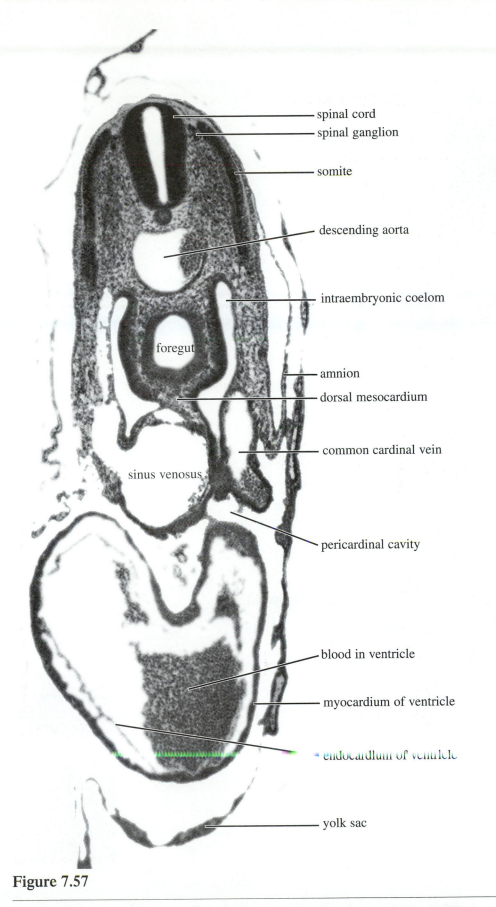

spinal cord
spinal ganglion

somite

descending aorta

intraembryonic coelom

foregut

amnion
dorsal mesocardium

common cardinal vein

sinus venosus

pericardinal cavity

blood in ventricle

myocardium of ventricle

endocardium of ventricle

yolk sac

Figure 7.57

2-day chick embryo (stage 15), transverse section through the sinus venosus (110X).

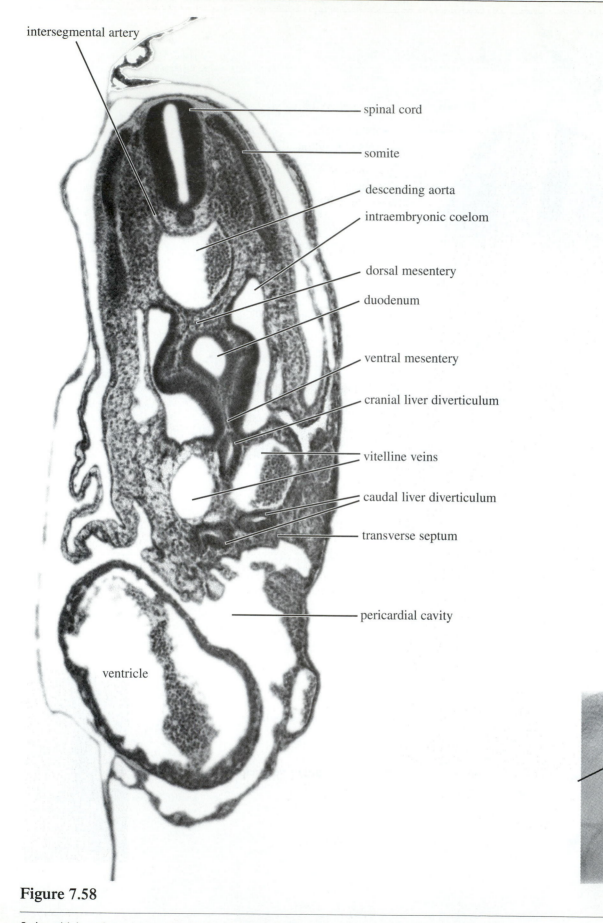

intersegmental artery

spinal cord

somite

descending aorta

intraembryonic coelom

dorsal mesentery

duodenum

ventral mesentery

cranial liver diverticulum

vitelline veins

caudal liver diverticulum

transverse septum

pericardial cavity

ventricle

Figure 7.58

2-day chick embryo (stage 15), transverse section through the liver rudiments (110X).

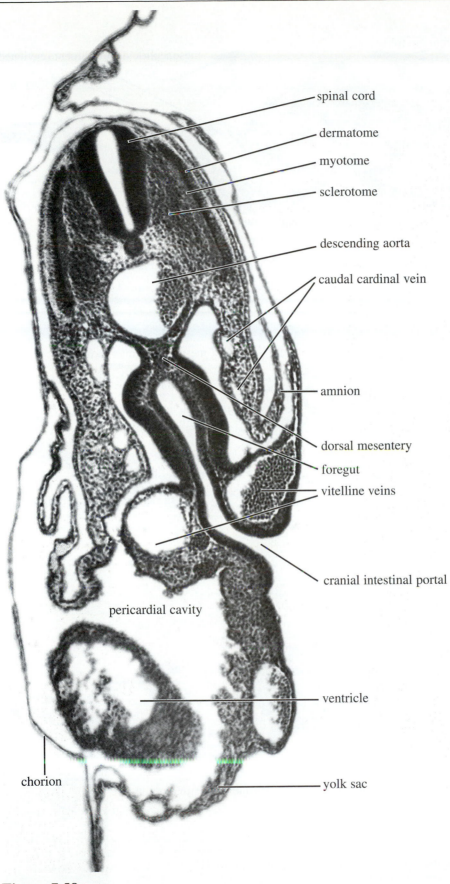

spinal cord

dermatome

myotome

sclerotome

descending aorta

caudal cardinal vein

amnion

dorsal mesentery

foregut

vitelline veins

cranial intestinal portal

pericardial cavity

ventricle

chorion

yolk sac

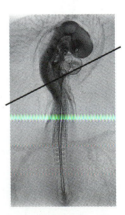

Figure 7.59

2-day chick embryo (stage 15), transverse section through the cranial intestinal portal (110X).

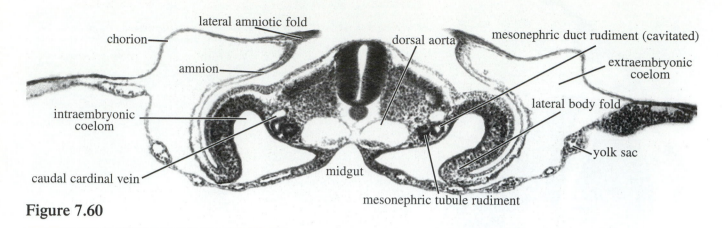

Figure 7.60

2-day chick embryo (stage 15), transverse section through the mesonephros (110X).

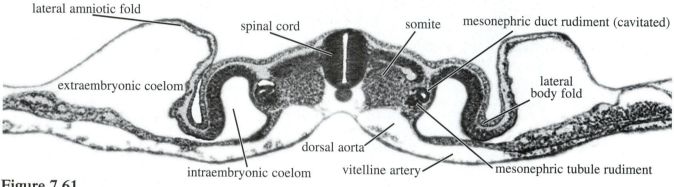

Figure 7.61

2-day chick embryo (stage 15), transverse section through the vitelline artery (110X).

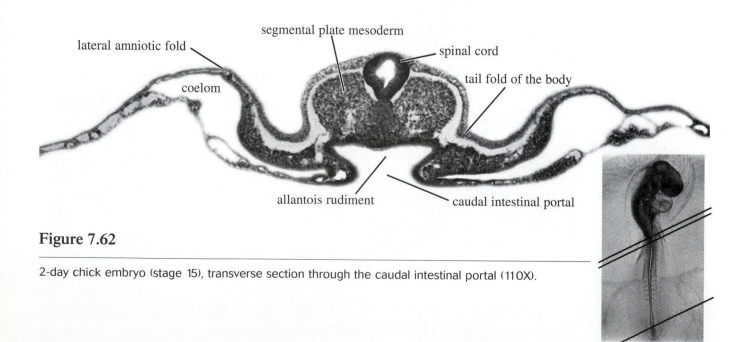

Figure 7.62

2-day chick embryo (stage 15), transverse section through the caudal intestinal portal (110X).

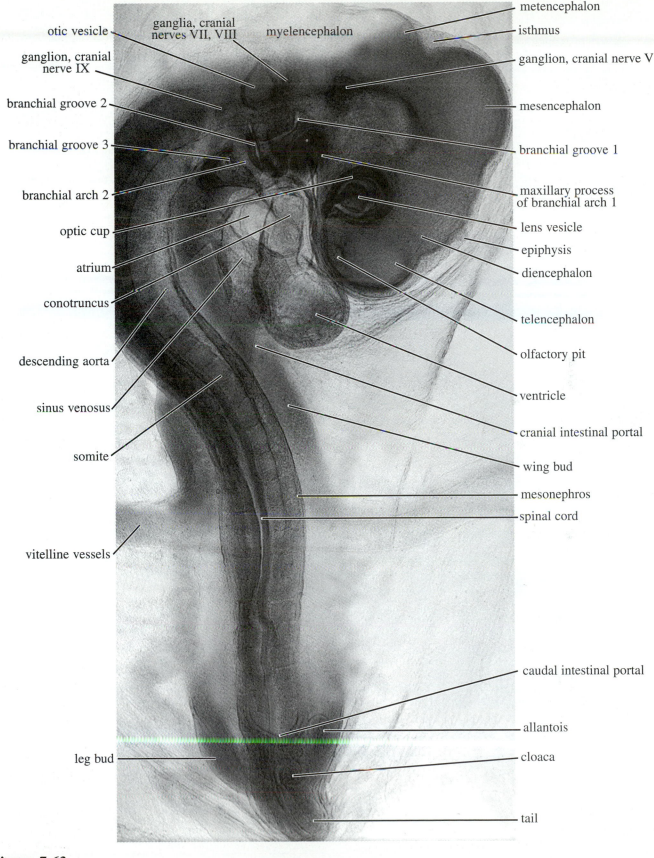

otic vesicle

ganglia, cranial
nerves VII, VIII

myelencephalon

metencephalon

isthmus

ganglion, cranial
nerve IX

ganglion, cranial nerve V

branchial groove 2

mesencephalon

branchial groove 3

branchial groove 1

branchial arch 2

maxillary process
of branchial arch 1

optic cup

lens vesicle

atrium

epiphysis

conotruncus

diencephalon

descending aorta

telencephalon

sinus venosus

olfactory pit

somite

ventricle

cranial intestinal portal

wing bud

vitelline vessels

mesonephros

spinal cord

caudal intestinal portal

allantois

leg bud

cloaca

tail

Figure 7.63

3-day chick embryo (stage 18), whole mount, transmitted illumination (30X).

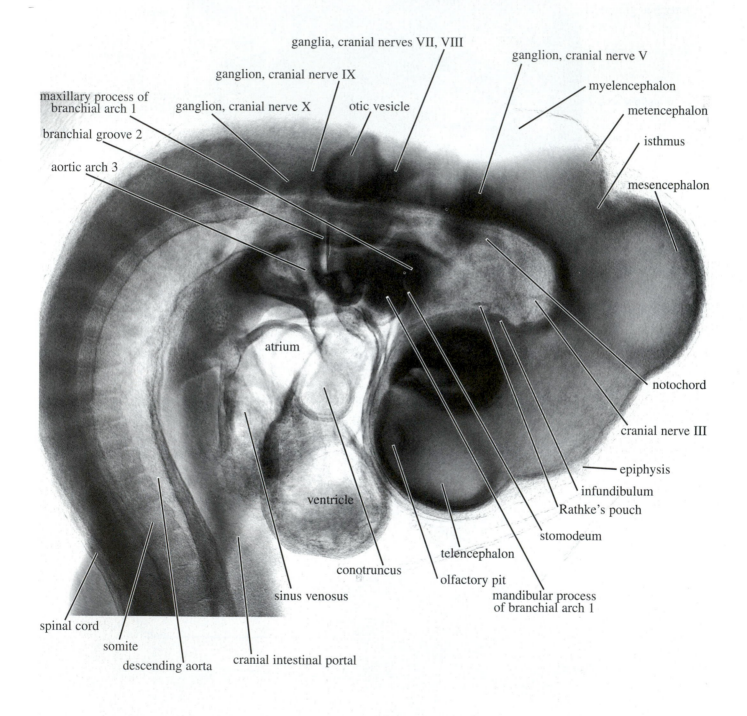

Figure 7.64

Cranial third of a 3-day chick embryo (stage 18), whole mount, transmitted illumination (65X).

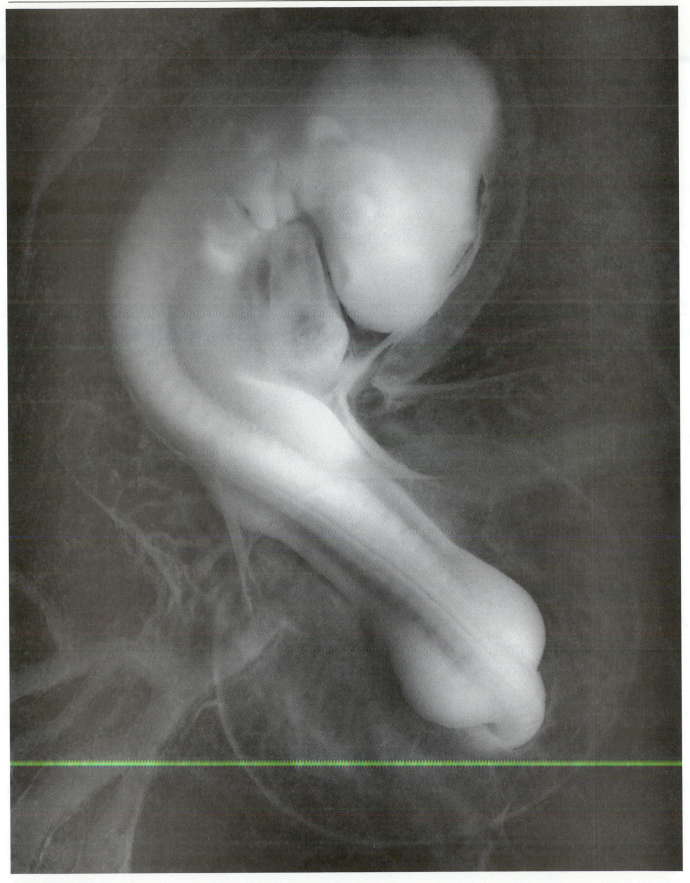

Figure 7.65

3-day chick embryo (stage 19), whole mount, incident illumination (40X).

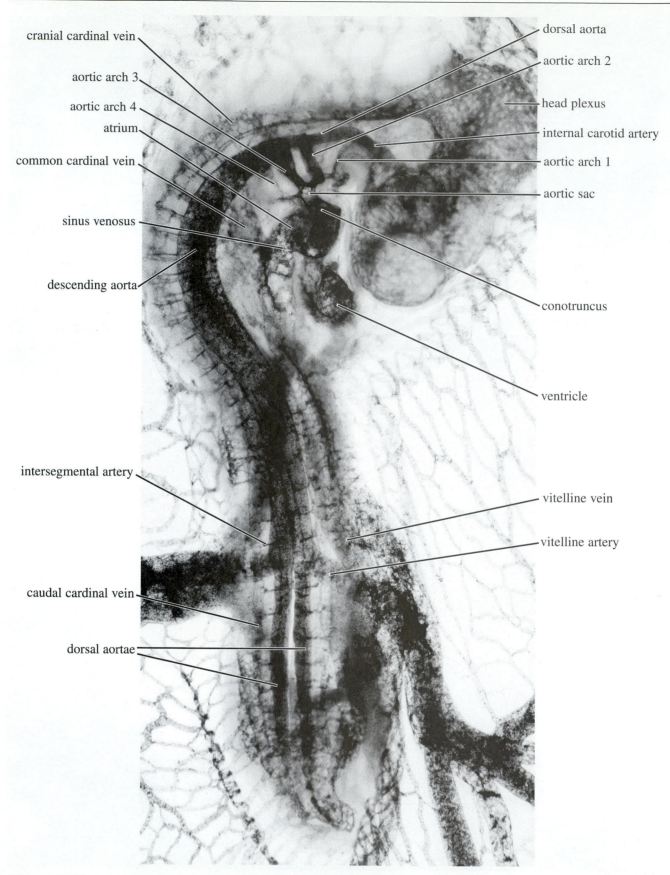

cranial cardinal vein

aortic arch 3

aortic arch 4

atrium

common cardinal vein

sinus venosus

descending aorta

intersegmental artery

caudal cardinal vein

dorsal aortae

dorsal aorta

aortic arch 2

head plexus

internal carotid artery

aortic arch 1

aortic sac

conotruncus

ventricle

vitelline vein

vitelline artery

Figure 7.66

3-day chick embryo (stage 18), whole mount, blood vessels injected with India ink (35X).

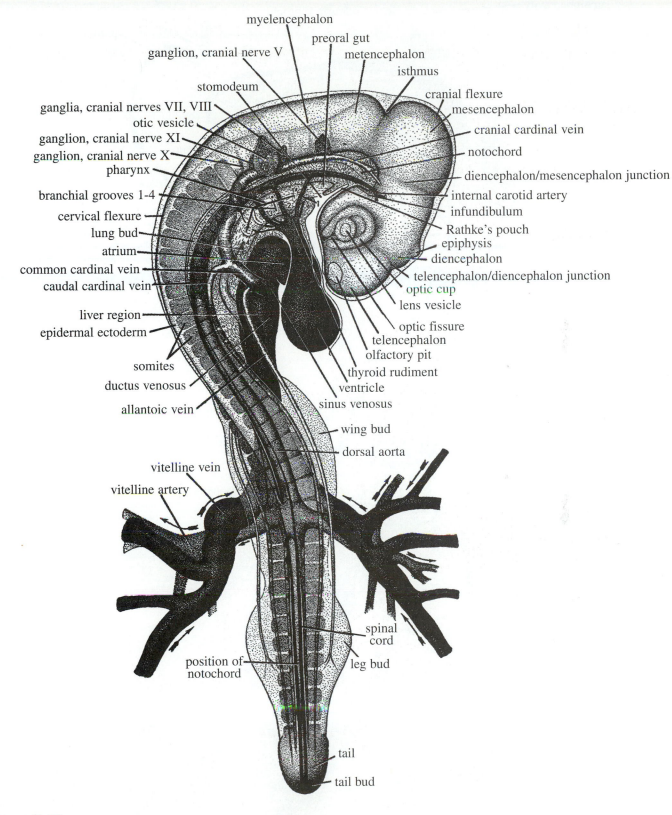

Figure 7.67

Drawing of a dorsal view of a whole mount of a 3-day chick embryo (stage 18) with 35 somites. Arrows indicate the directions of blood flow.

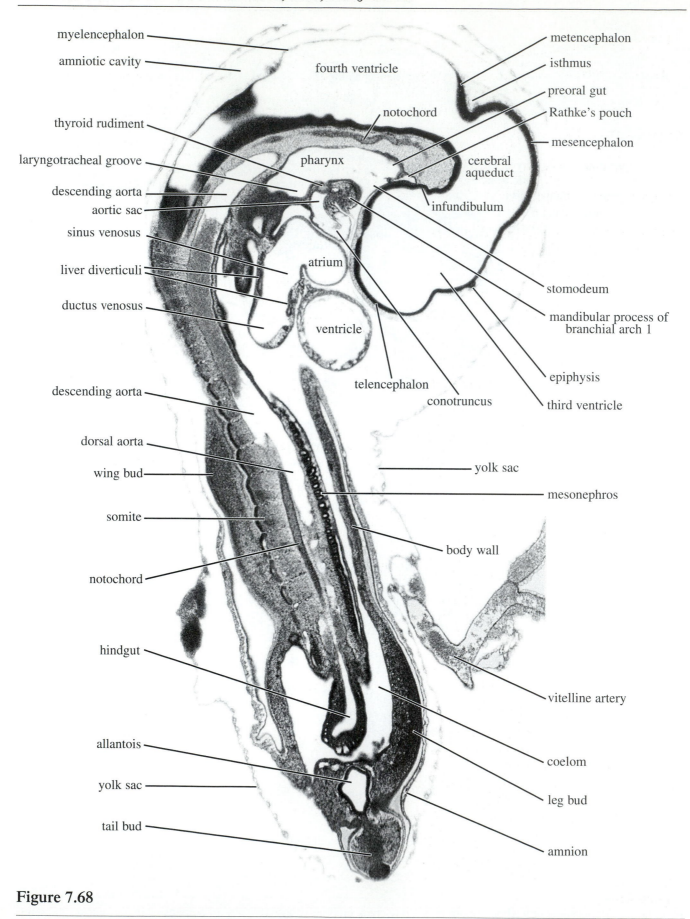

myelencephalon
amniotic cavity
thyroid rudiment
laryngotracheal groove
descending aorta
aortic sac
sinus venosus
liver diverticuli
ductus venosus
descending aorta
dorsal aorta
wing bud
somite
notochord
hindgut
allantois
yolk sac
tail bud

fourth ventricle
notochord
pharynx
atrium
ventricle
telencephalon
conotruncus

metencephalon
isthmus
preoral gut
Rathke's pouch
mesencephalon
cerebral aqueduct
infundibulum
stomodeum
mandibular process of branchial arch 1
epiphysis
third ventricle
yolk sac
mesonephros
body wall
vitelline artery
coelom
leg bud
amnion

Figure 7.68

3-day chick embryo (stage 18), midsagittal section (30X).

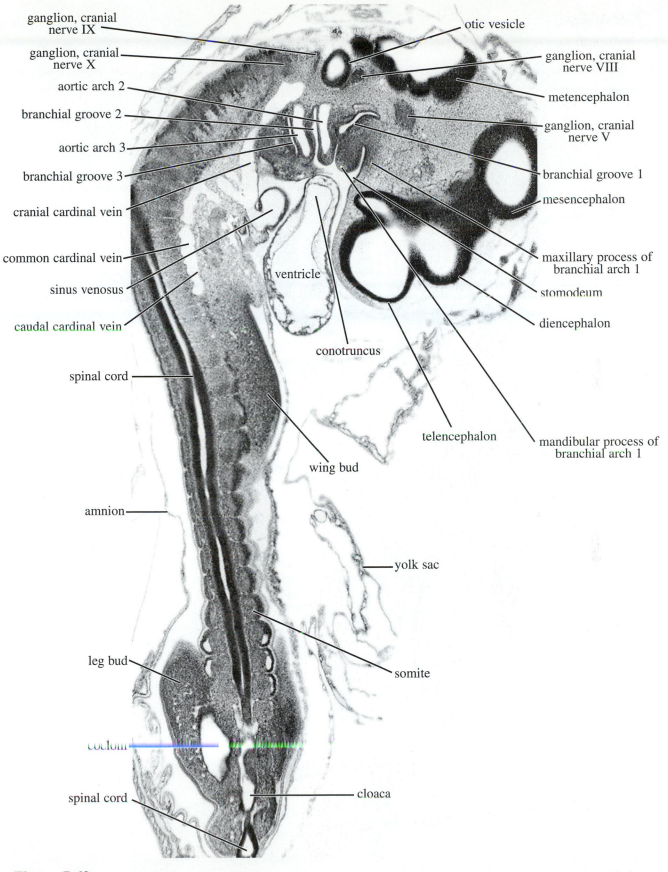

ganglion, cranial
nerve IX

ganglion, cranial
nerve X

aortic arch 2

branchial groove 2

aortic arch 3

branchial groove 3

cranial cardinal vein

common cardinal vein

sinus venosus

caudal cardinal vein

spinal cord

amnion

leg bud

coelom

spinal cord

otic vesicle

ganglion, cranial
nerve VIII

metencephalon

ganglion, cranial
nerve V

branchial groove 1

mesencephalon

maxillary process of
branchial arch 1

stomodeum

diencephalon

ventricle

conotruncus

wing bud

telencephalon

mandibular process of
branchial arch 1

yolk sac

somite

cloaca

Figure 7.69

3-day chick embryo (stage 18), parasagittal section, right side (30X).

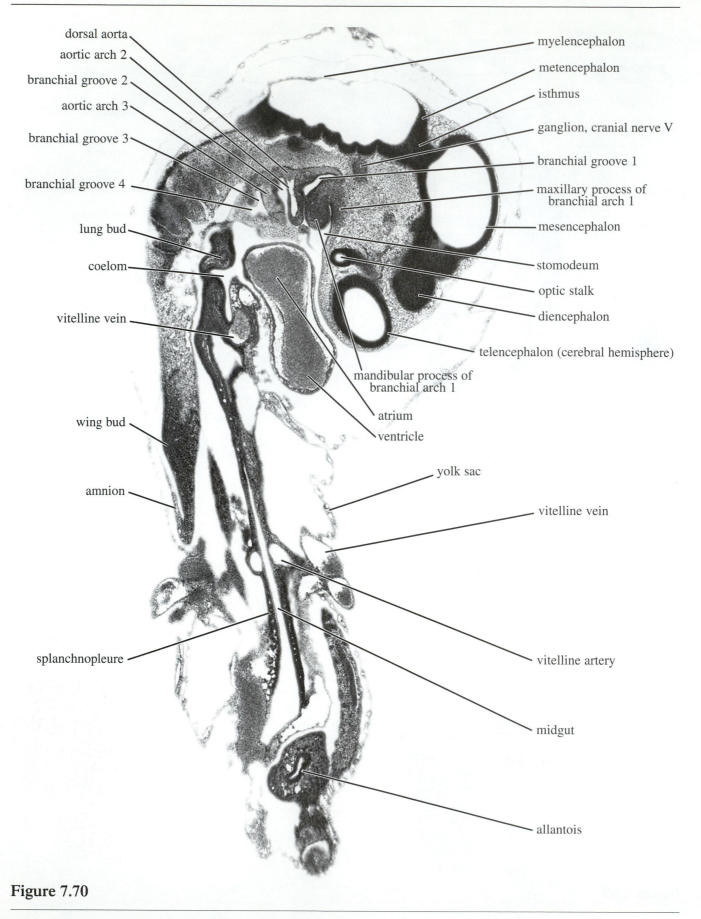

Figure 7.70

3-day chick embryo (stage 18), parasagittal section, left side (30X).

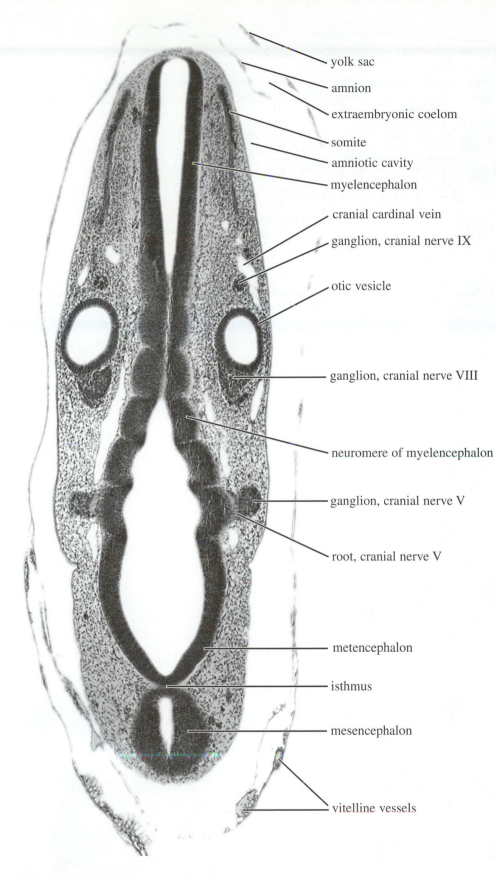

yolk sac

amnion

extraembryonic coelom

somite

amniotic cavity

myelencephalon

cranial cardinal vein

ganglion, cranial nerve IX

otic vesicle

ganglion, cranial nerve VIII

neuromere of myelencephalon

ganglion, cranial nerve V

root, cranial nerve V

metencephalon

isthmus

mesencephalon

vitelline vessels

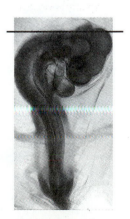

Figure 7.71

3-day chick embryo (stage 18), transverse section through the otic vesicles (65X).

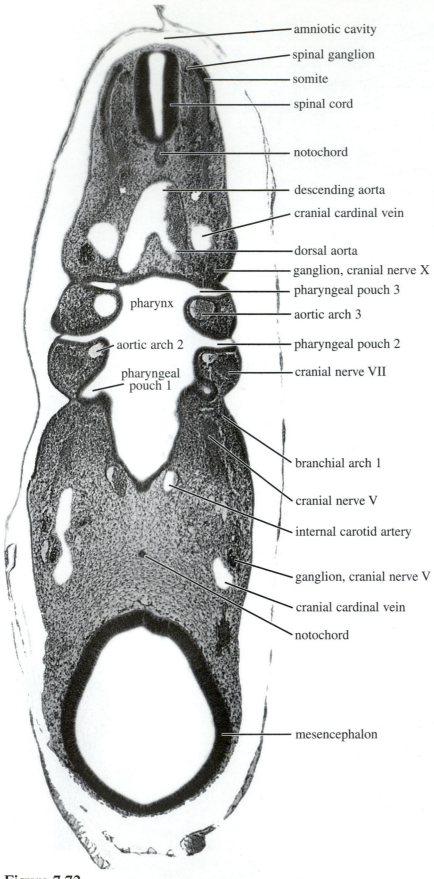

amniotic cavity

spinal ganglion

somite

spinal cord

notochord

descending aorta

cranial cardinal vein

dorsal aorta

ganglion, cranial nerve X

pharyngeal pouch 3

pharynx

aortic arch 3

aortic arch 2

pharyngeal pouch 2

pharyngeal pouch 1

cranial nerve VII

branchial arch 1

cranial nerve V

internal carotid artery

ganglion, cranial nerve V

cranial cardinal vein

notochord

mesencephalon

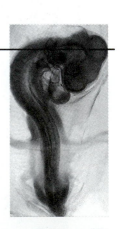

Figure 7.72

3-day chick embryo (stage 18), transverse section through the pharynx (65X).

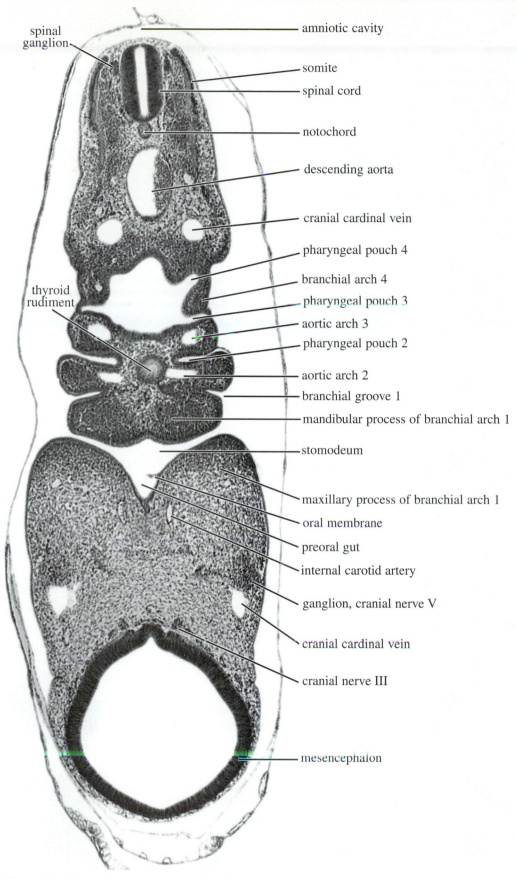

spinal ganglion

amniotic cavity

somite

spinal cord

notochord

descending aorta

cranial cardinal vein

pharyngeal pouch 4

branchial arch 4

pharyngeal pouch 3

aortic arch 3

pharyngeal pouch 2

aortic arch 2

branchial groove 1

mandibular process of branchial arch 1

stomodeum

maxillary process of branchial arch 1

oral membrane

preoral gut

internal carotid artery

ganglion, cranial nerve V

cranial cardinal vein

cranial nerve III

thyroid rudiment

mesencephalon

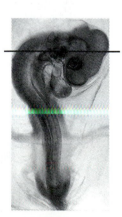

Figure 7.73

3-day chick embryo (stage 18), transverse section through the thyroid rudiment (65X).

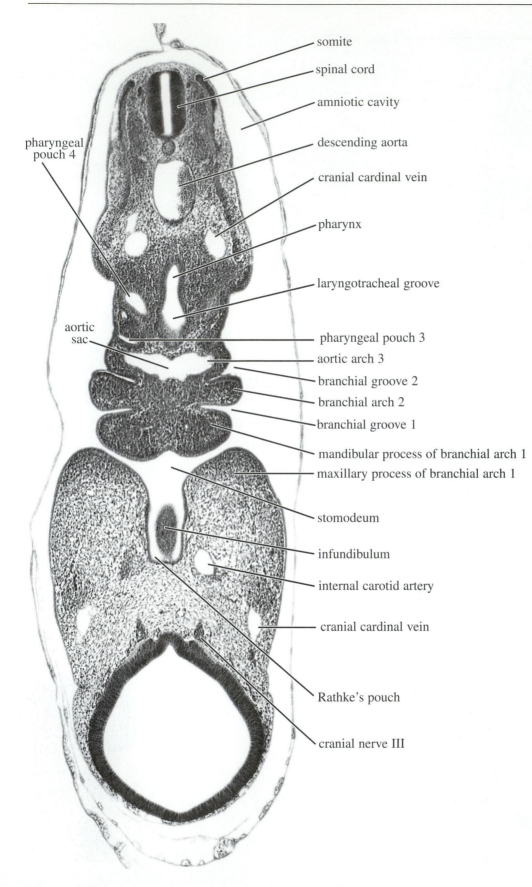

- somite
- spinal cord
- amniotic cavity
- descending aorta
- cranial cardinal vein
- pharynx
- laryngotracheal groove
- pharyngeal pouch 3
- aortic arch 3
- branchial groove 2
- branchial arch 2
- branchial groove 1
- mandibular process of branchial arch 1
- maxillary process of branchial arch 1
- stomodeum
- infundibulum
- internal carotid artery
- cranial cardinal vein
- Rathke's pouch
- cranial nerve III

pharyngeal pouch 4

aortic sac

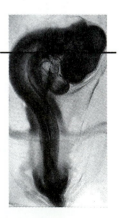

Figure 7.74

3-day chick embryo (stage 18), transverse section through the future hypophysis (65X).

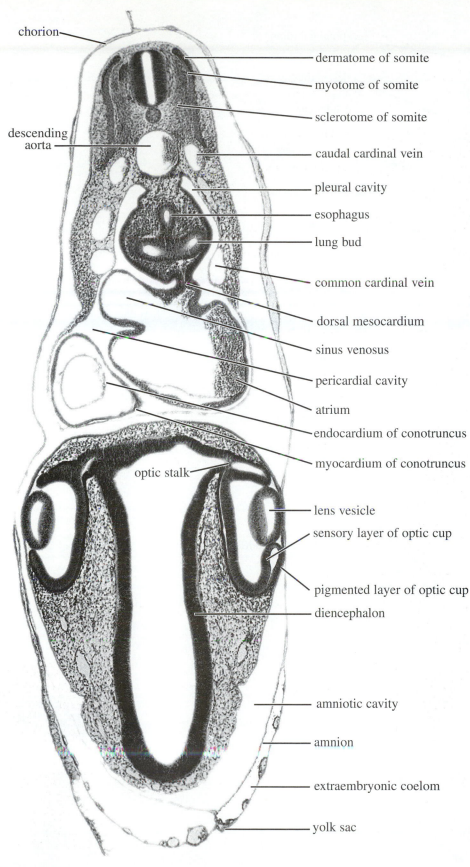

chorion

dermatome of somite

myotome of somite

sclerotome of somite

descending aorta

caudal cardinal vein

pleural cavity

esophagus

lung bud

common cardinal vein

dorsal mesocardium

sinus venosus

pericardial cavity

atrium

endocardium of conotruncus

myocardium of conotruncus

optic stalk

lens vesicle

sensory layer of optic cup

pigmented layer of optic cup

diencephalon

amniotic cavity

amnion

extraembryonic coelom

yolk sac

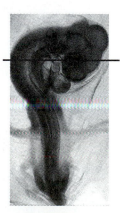

Figure 7.75

3-day chick embryo (stage 18), transverse section through the optic cups (65X).

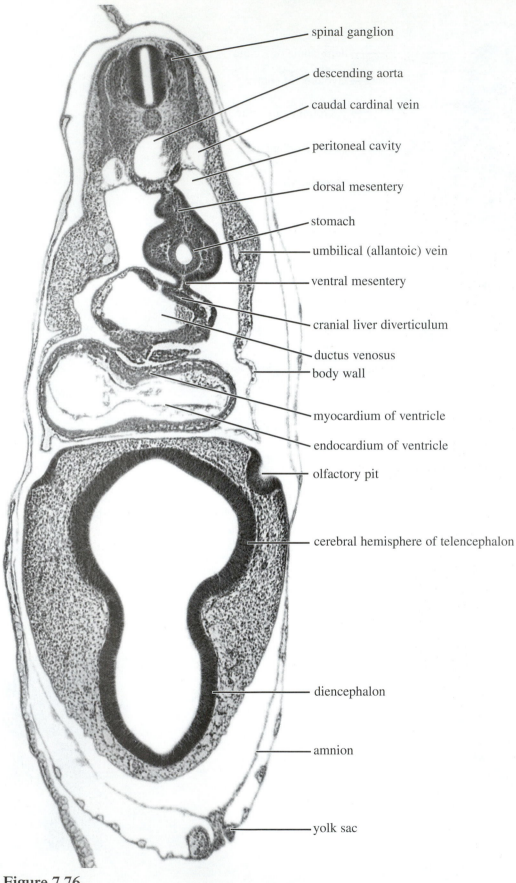

spinal ganglion

descending aorta

caudal cardinal vein

peritoneal cavity

dorsal mesentery

stomach

umbilical (allantoic) vein

ventral mesentery

cranial liver diverticulum

ductus venosus

body wall

myocardium of ventricle

endocardium of ventricle

olfactory pit

cerebral hemisphere of telencephalon

diencephalon

amnion

yolk sac

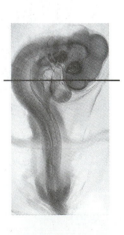

Figure 7.76

3-day chick embryo (stage 18), transverse section through the olfactory pits (65X).

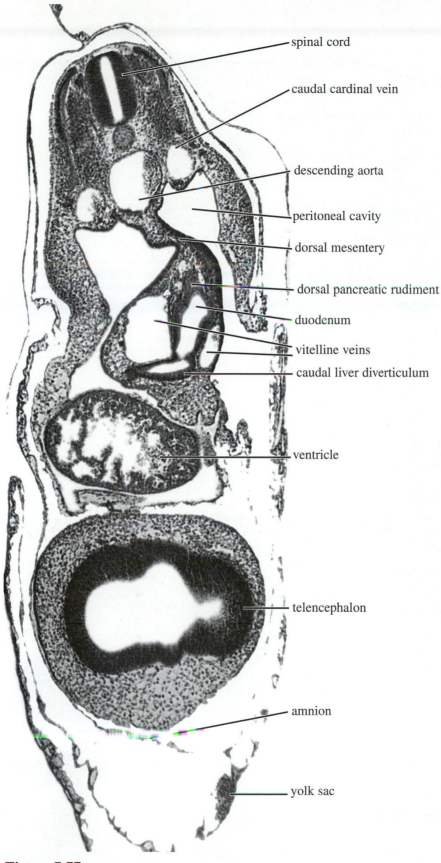

spinal cord

caudal cardinal vein

descending aorta

peritoneal cavity

dorsal mesentery

dorsal pancreatic rudiment

duodenum

vitelline veins

caudal liver diverticulum

ventricle

telencephalon

amnion

yolk sac

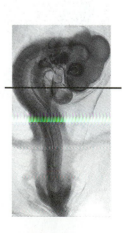

Figure 7.77

3-day chick embryo (stage 18), transverse section through the rudiments of the liver and pancreas (65X).

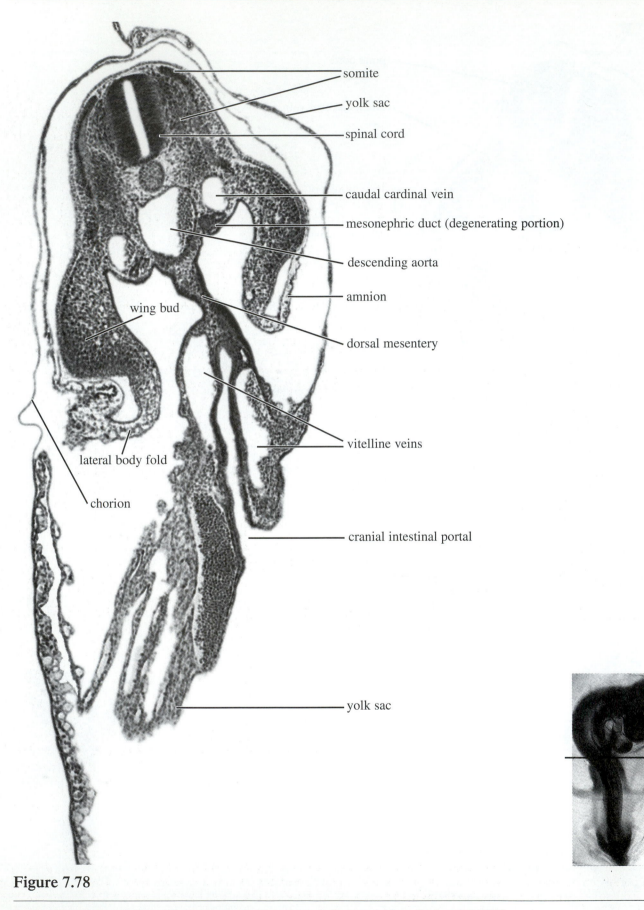

somite

yolk sac

spinal cord

caudal cardinal vein

mesonephric duct (degenerating portion)

descending aorta

amnion

dorsal mesentery

wing bud

vitelline veins

lateral body fold

chorion

cranial intestinal portal

yolk sac

Figure 7.78

3-day chick embryo (stage 18), transverse section through the cranial intestinal portal (65X).

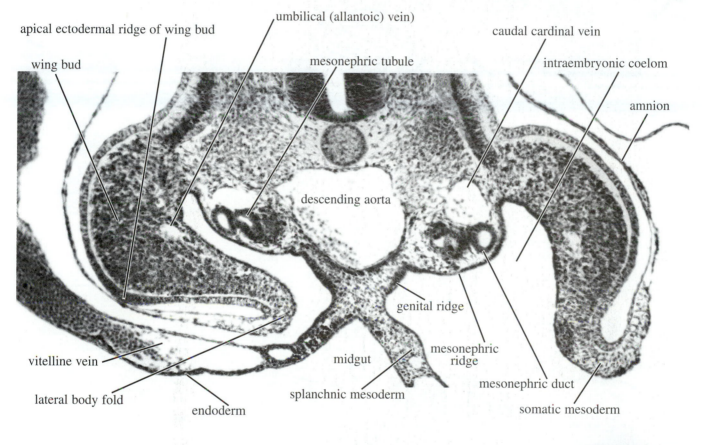

apical ectodermal ridge of wing bud

wing bud

umbilical (allantoic) vein)

mesonephric tubule

caudal cardinal vein

intraembryonic coelom

amnion

descending aorta

genital ridge

vitelline vein

midgut

mesonephric ridge

mesonephric duct

lateral body fold

endoderm

splanchnic mesoderm

somatic mesoderm

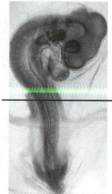

Figure 7.79

3-day chick embryo (stage 18), central portion of a transverse section through the genital ridge (150X).

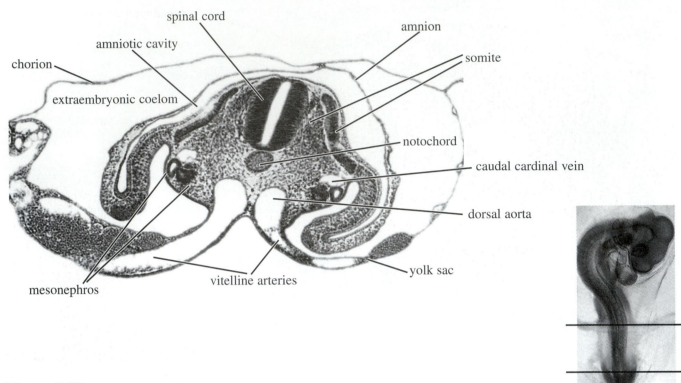

Figure 7.80

3-day chick embryo (stage 18), transverse section through the vitelline arteries (80X).

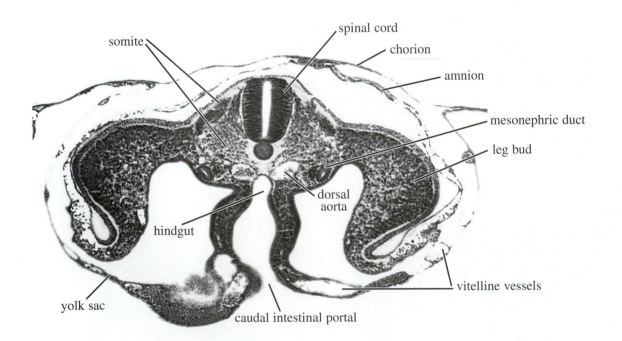

Figure 7.81

3-day chick embryo (stage 18), transverse section through the caudal intestinal portal (80X).

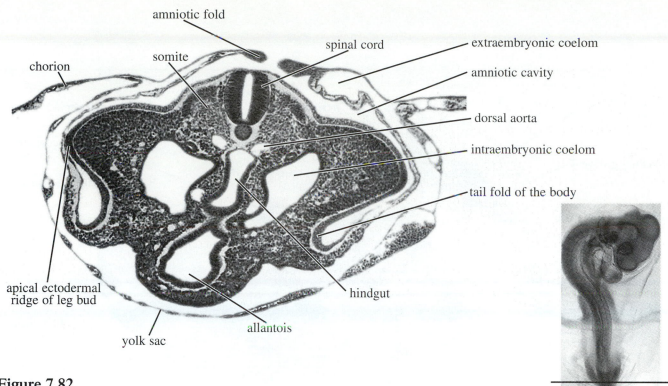

Figure 7.82

3-day chick embryo (stage 18), transverse section through the allantois (80X).

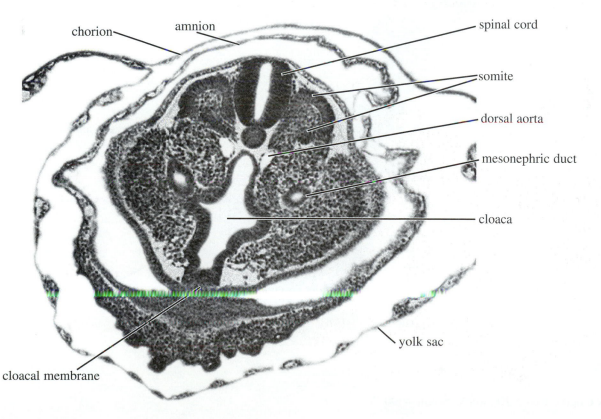

Figure 7.83

3-day chick embryo (stage 18), transverse section through the cloaca (80X).

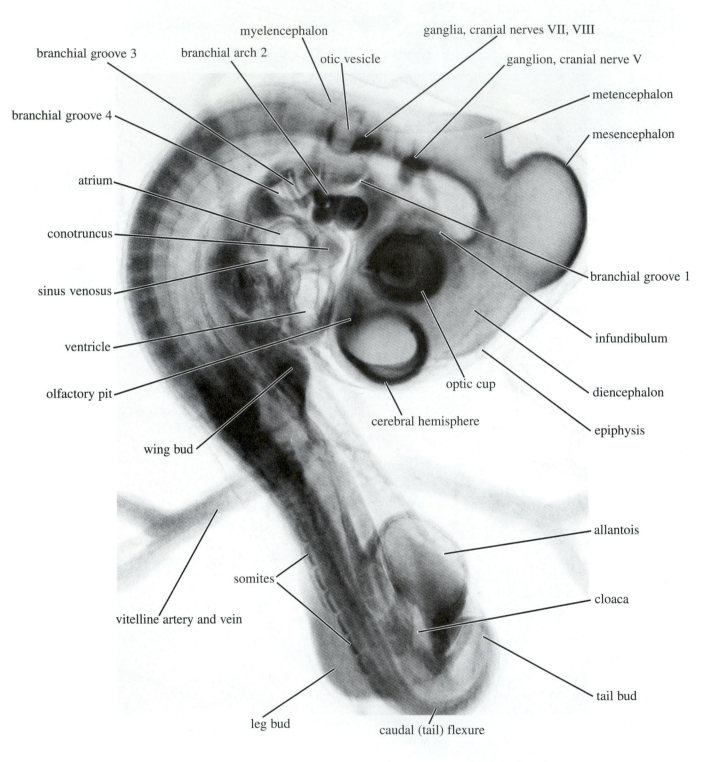

Figure 7.84

3.5-day chick embryo (stage 20), whole mount (25X).

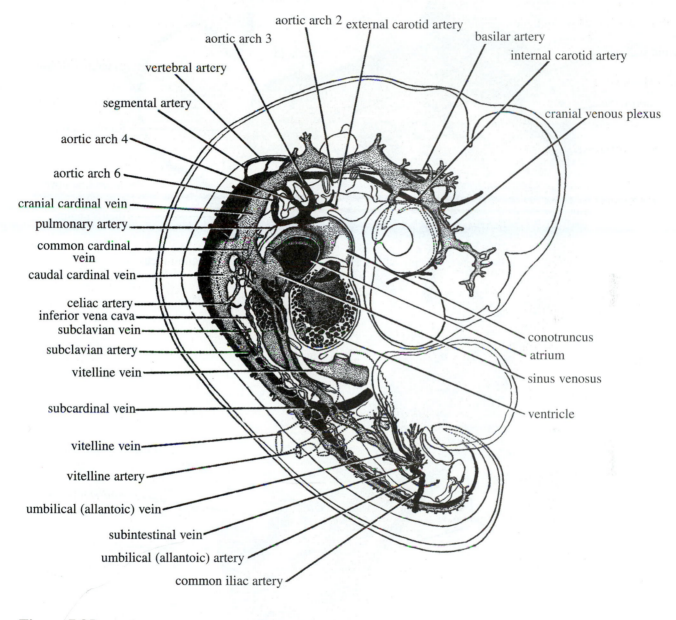

Figure 7.85

Reconstruction of the circulatory system of the 3.5-day chick embryo (stage 21), whole mount (18X).

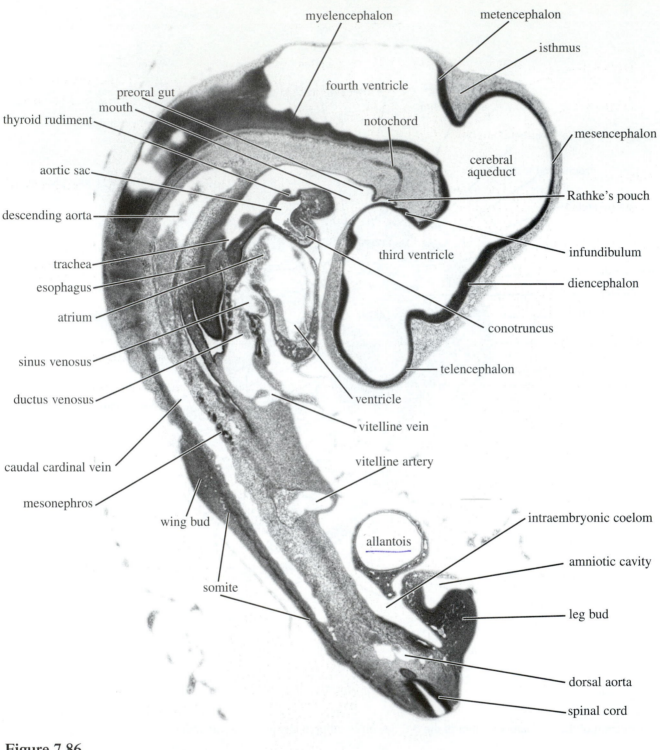

Figure 7.86

3.5-day chick embryo (stage 21), sagittal section (30X).

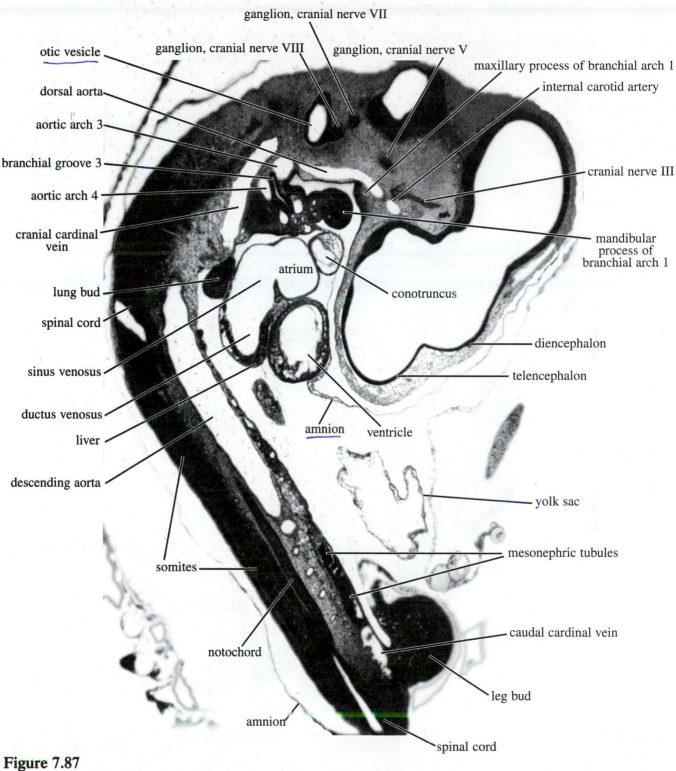

ganglion, cranial nerve VII

otic vesicle

ganglion, cranial nerve VIII

ganglion, cranial nerve V

maxillary process of branchial arch 1

dorsal aorta

internal carotid artery

aortic arch 3

branchial groove 3

aortic arch 4

cranial nerve III

cranial cardinal vein

mandibular process of branchial arch 1

atrium

conotruncus

lung bud

spinal cord

diencephalon

sinus venosus

telencephalon

ductus venosus

liver

amnion

ventricle

descending aorta

yolk sac

mesonephric tubules

somites

caudal cardinal vein

notochord

leg bud

amnion

spinal cord

Figure 7.87

3.5-day chick embryo (stage 21), sagittal section (30X).

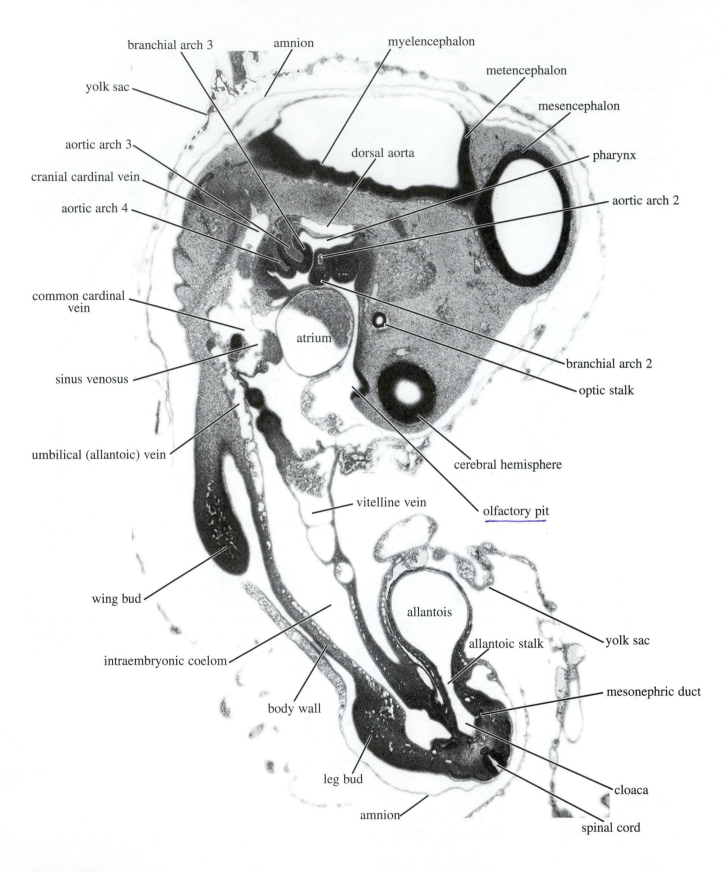

Figure 7.88

3.5-day chick embryo (stage 21), sagittal section (30X).

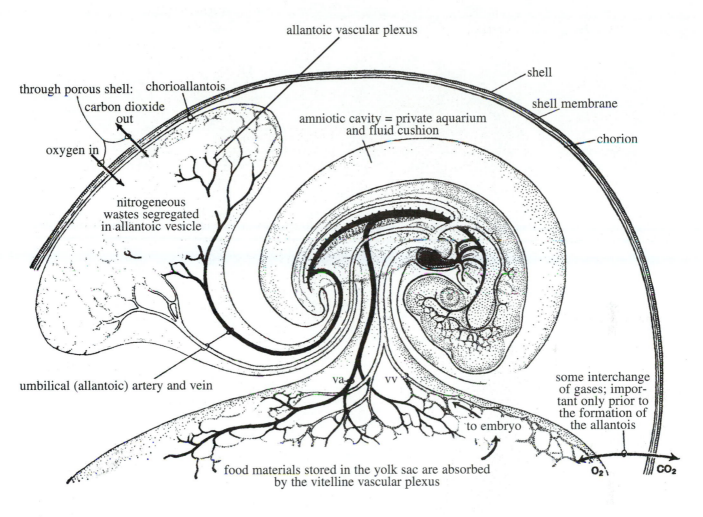

allantoic vascular plexus

through porous shell: chorioallantois

carbon dioxide
out

oxygen in

nitrogeneous
wastes segregated
in allantoic vesicle

shell

shell membrane

chorion

amniotic cavity = private aquarium
and fluid cushion

umbilical (allantoic) artery and vein

some interchange
of gases; impor-
tant only prior to
the formation of
the allantois

va   vv

to embryo

O_2 CO_2

food materials stored in the yolk sac are absorbed
by the vitelline vascular plexus

Figure 7.89

Schematic diagram showing the arrangement of the main circulatory channels in a 4-day chick embryo. The sites of some of the extraembryonic interchanges important in its physiology are indicated. The vessels within the embryo carry food and oxygen to all its growing tissues, and relieve them of the waste products incident to their metabolism. va, vitelline artery; vv, vitelline vein.

Mammalian Development

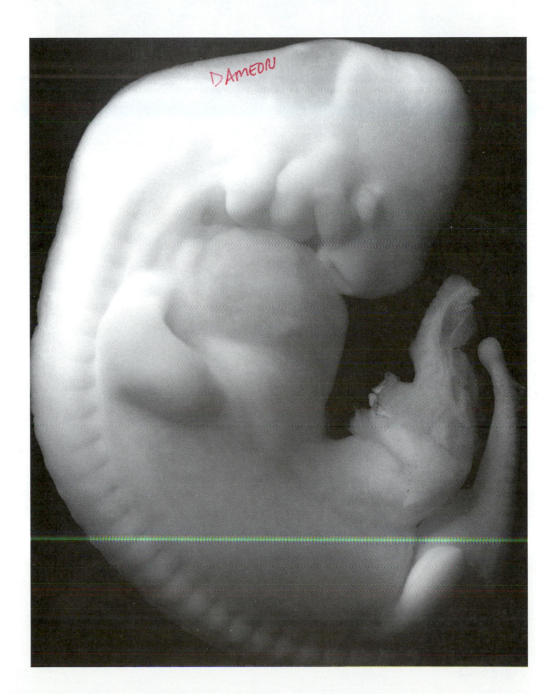

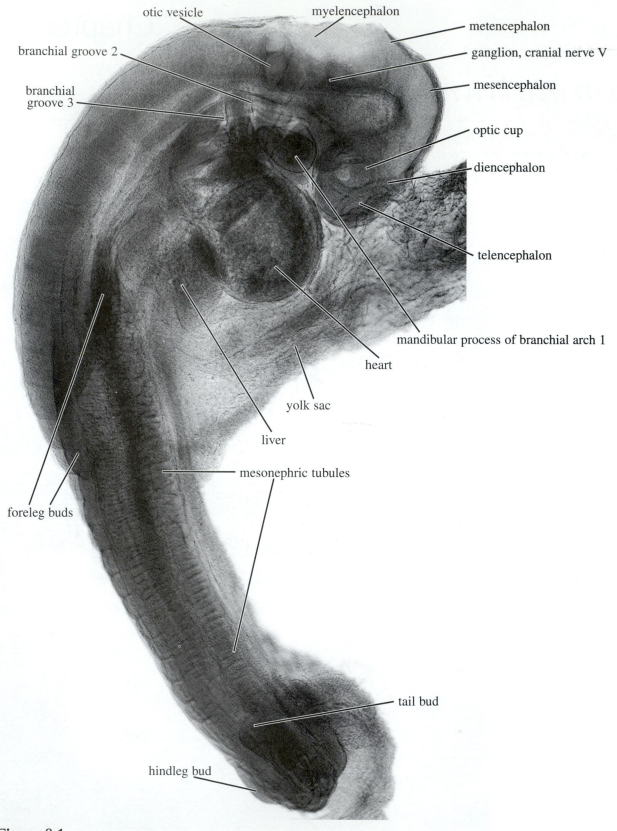

otic vesicle

myelencephalon

metencephalon

ganglion, cranial nerve V

branchial groove 2

mesencephalon

branchial
groove 3

optic cup

diencephalon

telencephalon

mandibular process of branchial arch 1

heart

yolk sac

liver

mesonephric tubules

foreleg buds

tail bud

hindleg bud

Figure 8.1

5-mm pig embryo, whole mount (30X).

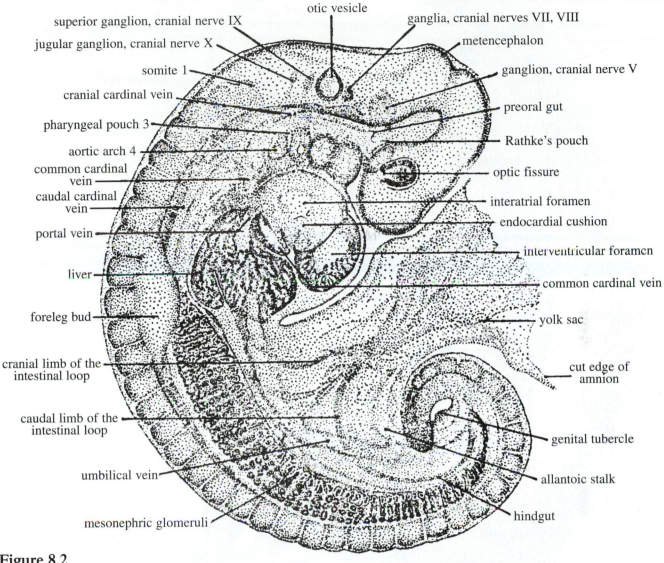

superior ganglion, cranial nerve IX

jugular ganglion, cranial nerve X

somite 1

cranial cardinal vein

pharyngeal pouch 3

aortic arch 4

common cardinal vein

caudal cardinal vein

portal vein

liver

foreleg bud

cranial limb of the intestinal loop

caudal limb of the intestinal loop

umbilical vein

mesonephric glomeruli

otic vesicle

ganglia, cranial nerves VII, VIII

metencephalon

ganglion, cranial nerve V

preoral gut

Rathke's pouch

optic fissure

interatrial foramen

endocardial cushion

interventricular foramen

common cardinal vein

yolk sac

cut edge of amnion

genital tubercle

allantoic stalk

hindgut

Figure 8.2

Drawing of a 5-mm pig embryo, whole mount (17X).

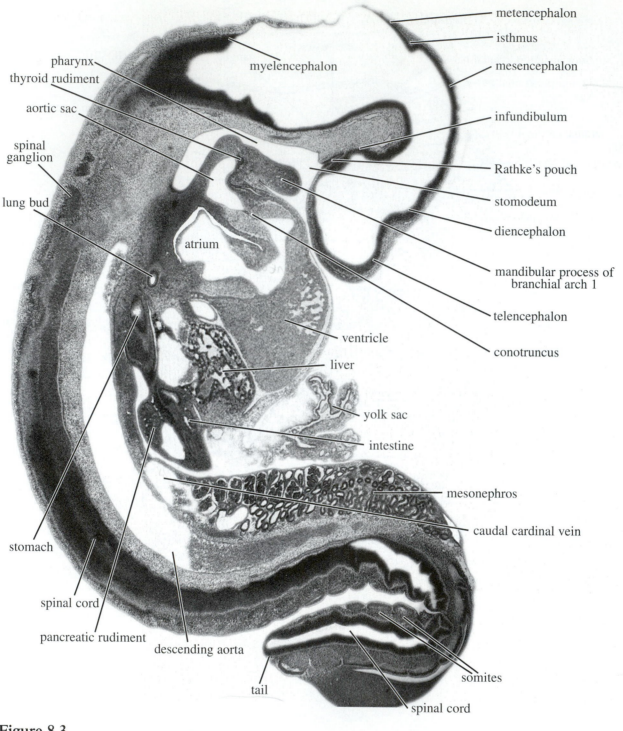

metencephalon

isthmus

mesencephalon

infundibulum

Rathke's pouch

stomodeum

diencephalon

mandibular process of
branchial arch 1

telencephalon

conotruncus

myelencephalon

pharynx

thyroid rudiment

aortic sac

spinal
ganglion

lung bud

atrium

ventricle

liver

yolk sac

intestine

mesonephros

caudal cardinal vein

stomach

spinal cord

pancreatic rudiment

descending aorta

tail

somites

spinal cord

Figure 8.3

6-mm pig embryo, sagittal section (30X).

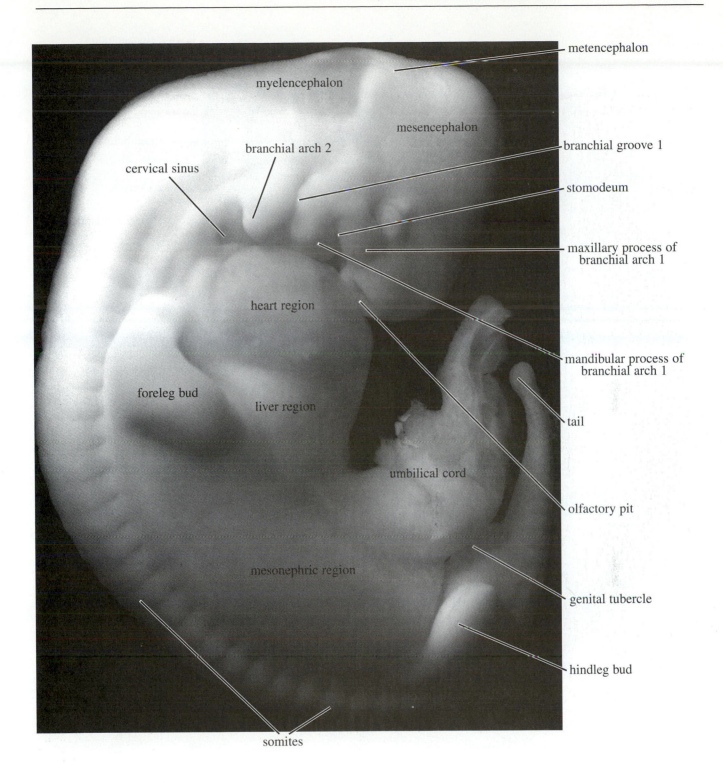

metencephalon

myelencephalon

mesencephalon

branchial arch 2

branchial groove 1

cervical sinus

stomodeum

maxillary process of branchial arch 1

heart region

mandibular process of branchial arch 1

foreleg bud

liver region

tail

umbilical cord

olfactory pit

mesonephric region

genital tubercle

hindleg bud

somites

Figure 8.4

10-mm pig embryo, opaque whole mount (20X; incident illumination).

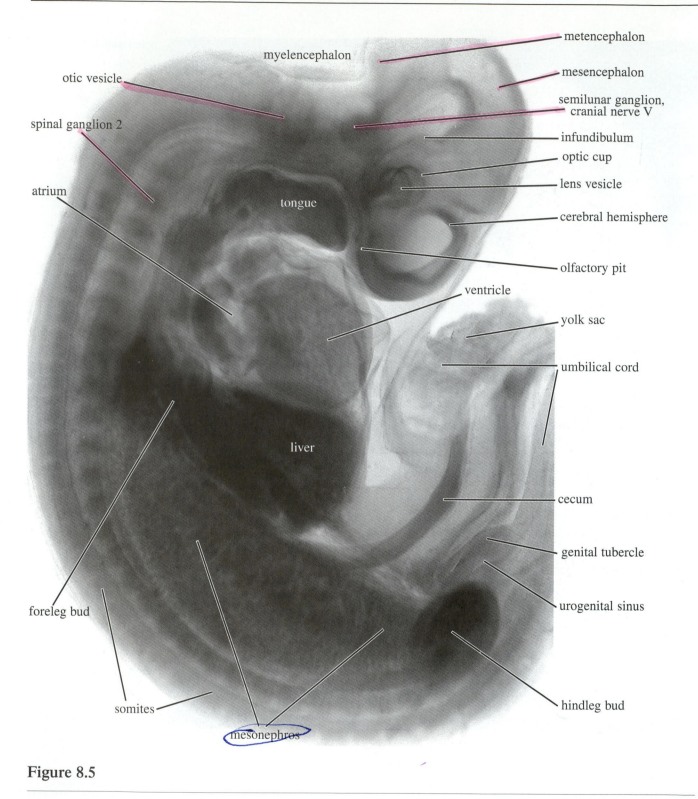

Figure 8.5

10-mm pig embryo, whole mount (20X; transmitted illumination).

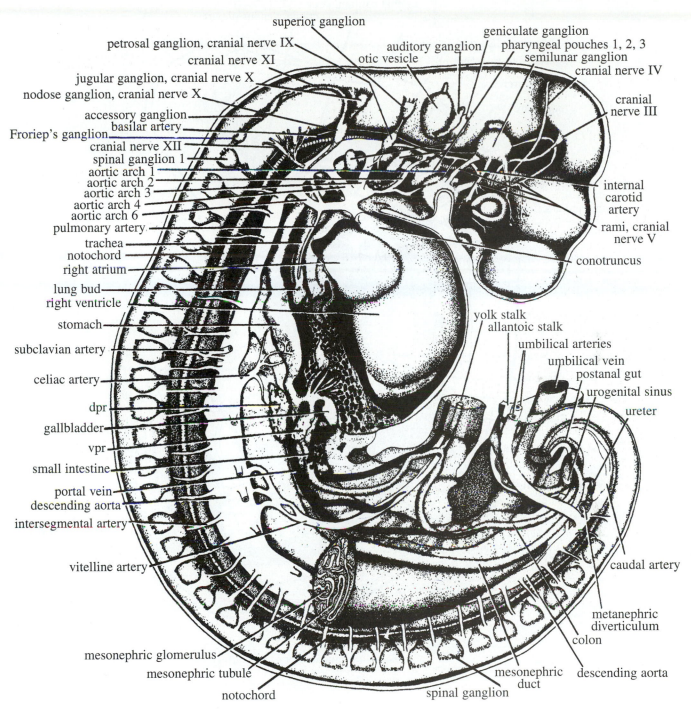

Figure 8.6

Reconstruction of the 10-mm pig embryo. dpr, dorsal pancreatic rudiment; vpr, ventral pancreatic rudiment.

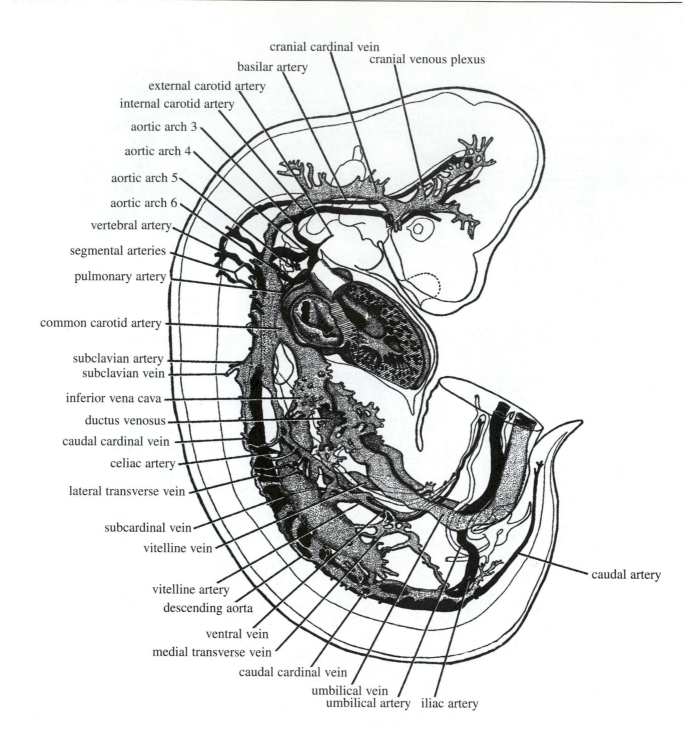

Figure 8.7

Reconstruction of the circulatory system of a 9.4-mm pig embryo (14X).

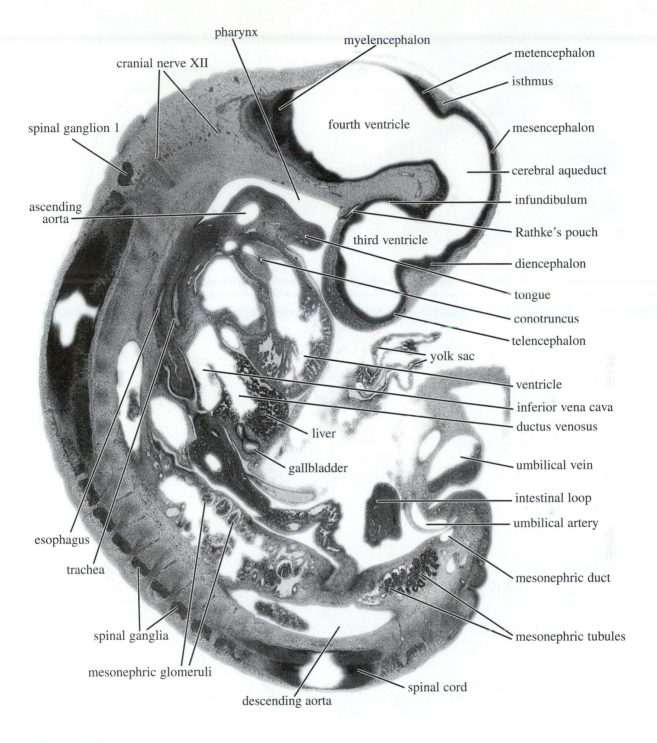

Figure 8.8

10-mm pig embryo, sagittal section (20X).

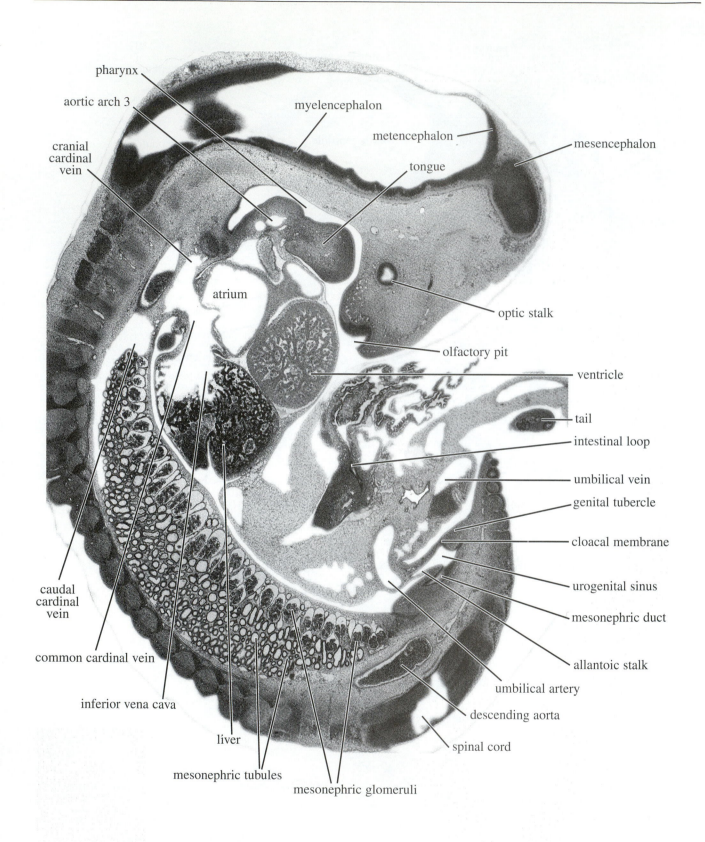

Figure 8.9

10-mm pig embryo, parasagittal section (20X).

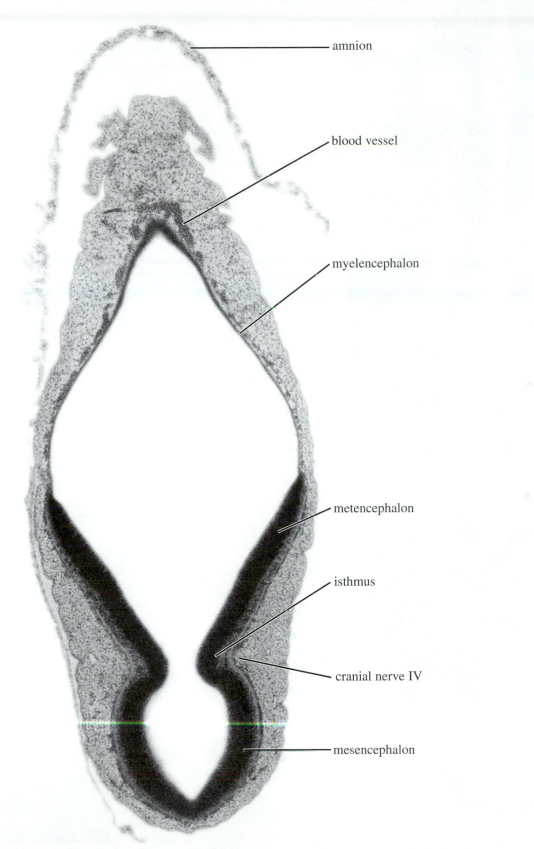

amnion

blood vessel

myelencephalon

metencephalon

isthmus

cranial nerve IV

mesencephalon

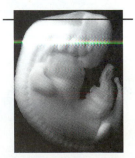

Figure 8.10

10-mm pig embryo, transverse section through cranial nerve IV (60X).

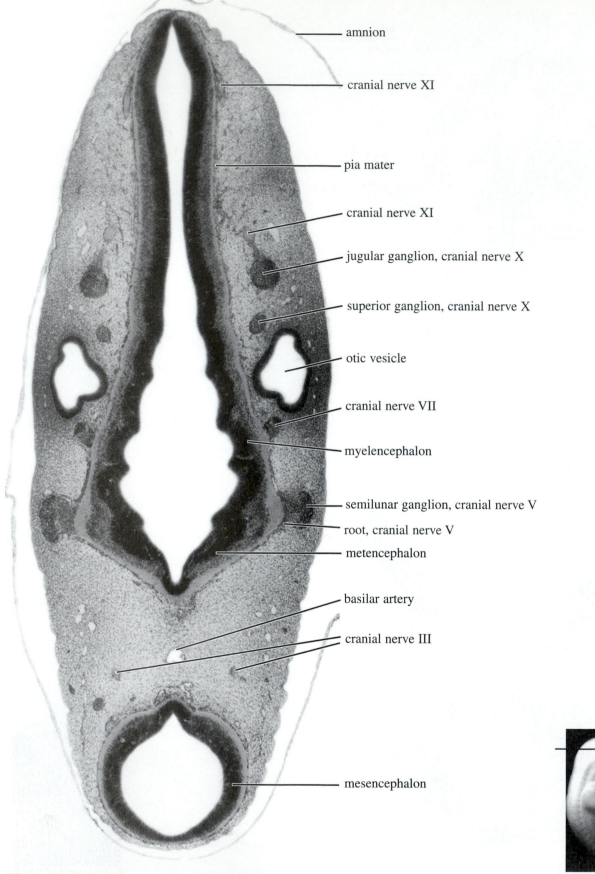

amnion

cranial nerve XI

pia mater

cranial nerve XI

jugular ganglion, cranial nerve X

superior ganglion, cranial nerve X

otic vesicle

cranial nerve VII

myelencephalon

semilunar ganglion, cranial nerve V

root, cranial nerve V

metencephalon

basilar artery

cranial nerve III

mesencephalon

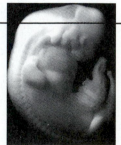

Figure 8.11

10-mm pig embryo, transverse section through the jugular and superior cranial ganglia (40X).

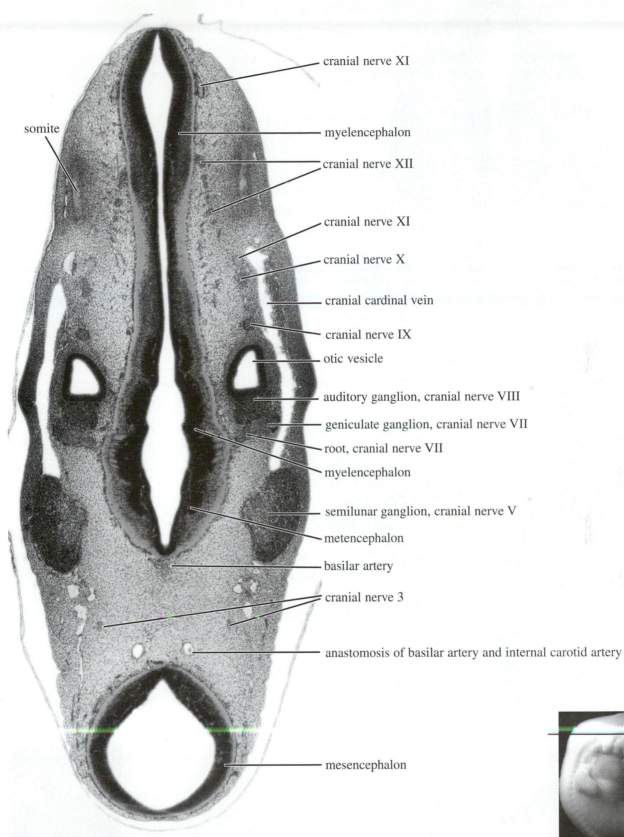

cranial nerve XI

myelencephalon

cranial nerve XII

cranial nerve XI

cranial nerve X

cranial cardinal vein

cranial nerve IX

otic vesicle

auditory ganglion, cranial nerve VIII

geniculate ganglion, cranial nerve VII

root, cranial nerve VII

myelencephalon

semilunar ganglion, cranial nerve V

metencephalon

basilar artery

cranial nerve 3

anastomosis of basilar artery and internal carotid artery

mesencephalon

somite

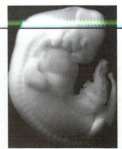

Figure 8.12

10-mm pig embryo, transverse section through the semilunar and geniculate cranial ganglia (40X).

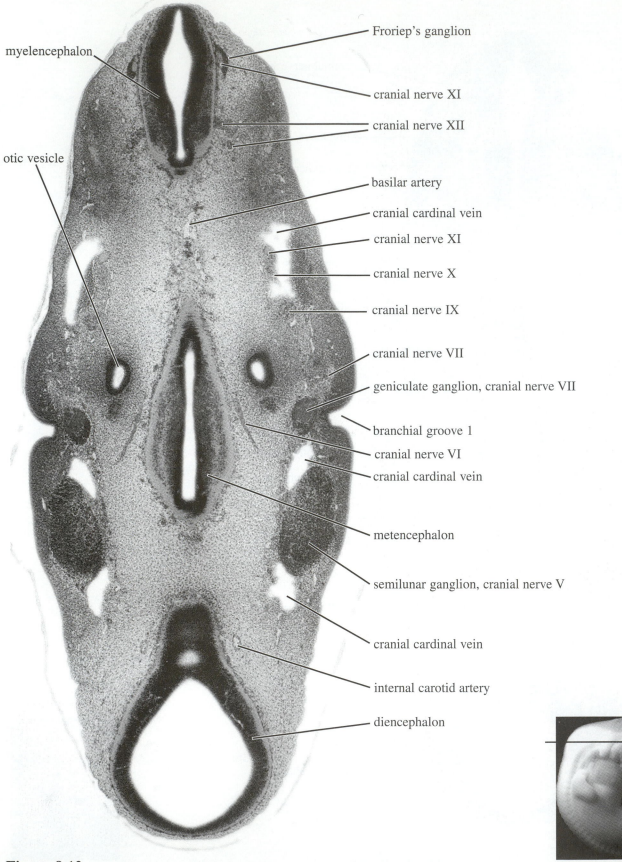

myelencephalon

Froriep's ganglion

cranial nerve XI

cranial nerve XII

otic vesicle

basilar artery

cranial cardinal vein

cranial nerve XI

cranial nerve X

cranial nerve IX

cranial nerve VII

geniculate ganglion, cranial nerve VII

branchial groove 1

cranial nerve VI

cranial cardinal vein

metencephalon

semilunar ganglion, cranial nerve V

cranial cardinal vein

internal carotid artery

diencephalon

Figure 8.13

10-mm pig embryo, transverse section through cranial nerve VI (40X).

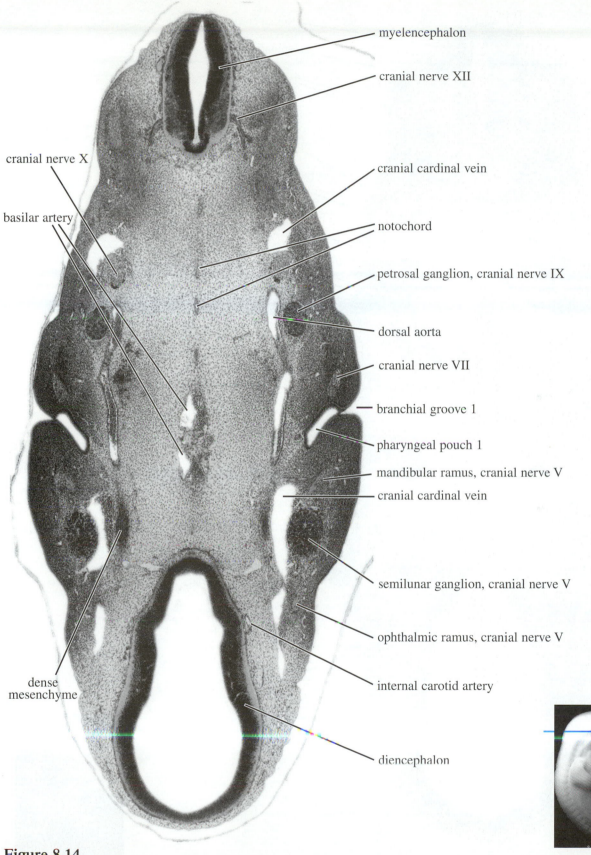

myelencephalon

cranial nerve XII

cranial cardinal vein

notochord

petrosal ganglion, cranial nerve IX

dorsal aorta

cranial nerve VII

branchial groove 1

pharyngeal pouch 1

mandibular ramus, cranial nerve V

cranial cardinal vein

semilunar ganglion, cranial nerve V

ophthalmic ramus, cranial nerve V

internal carotid artery

diencephalon

cranial nerve X

basilar artery

dense
mesenchyme

Figure 8.14

10-mm pig embryo, transverse section through pharyngeal pouch 1 (40X).

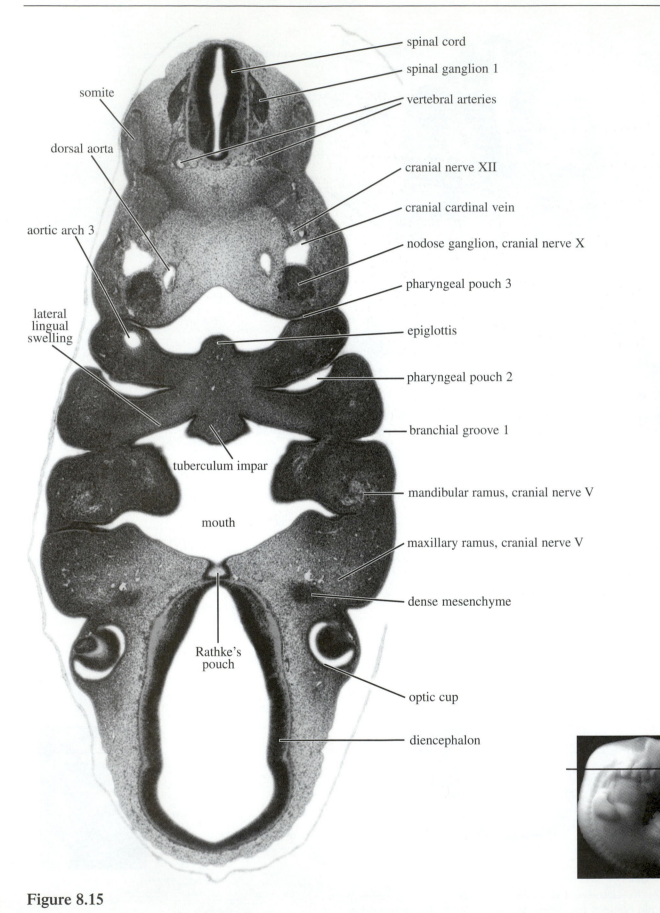

spinal cord

spinal ganglion 1

somite

vertebral arteries

dorsal aorta

cranial nerve XII

aortic arch 3

cranial cardinal vein

nodose ganglion, cranial nerve X

pharyngeal pouch 3

lateral lingual swelling

epiglottis

pharyngeal pouch 2

branchial groove 1

tuberculum impar

mandibular ramus, cranial nerve V

mouth

maxillary ramus, cranial nerve V

dense mesenchyme

Rathke's pouch

optic cup

diencephalon

Figure 8.15

10-mm pig embryo, transverse section through Rathke's pouch (40X).

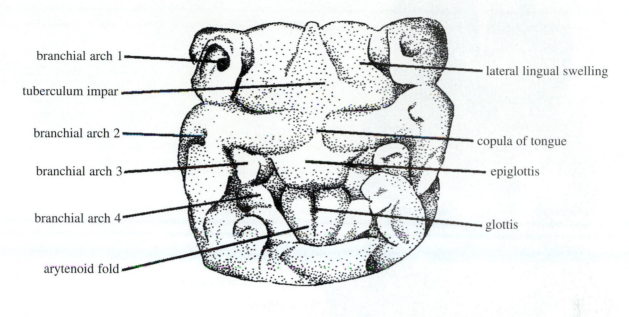

branchial arch 1

tuberculum impar

branchial arch 2

branchial arch 3

branchial arch 4

arytenoid fold

lateral lingual swelling

copula of tongue

epiglottis

glottis

Figure 8.16

Drawing of the floor of the mouth and pharynx of a 10-mm pig embryo, with the rest of the head removed.

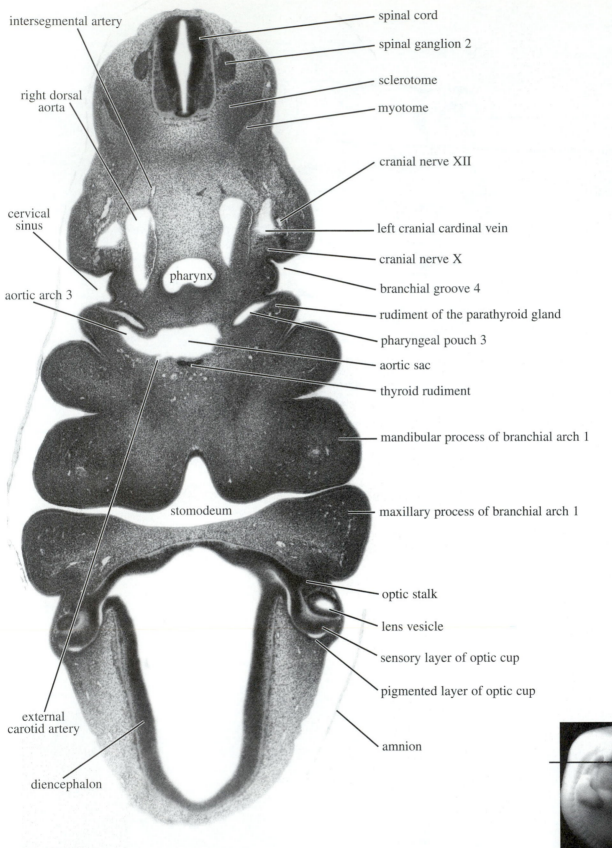

intersegmental artery

spinal cord

spinal ganglion 2

sclerotome

myotome

right dorsal aorta

cranial nerve XII

cervical sinus

left cranial cardinal vein

cranial nerve X

pharynx

branchial groove 4

aortic arch 3

rudiment of the parathyroid gland

pharyngeal pouch 3

aortic sac

thyroid rudiment

mandibular process of branchial arch 1

maxillary process of branchial arch 1

stomodeum

optic stalk

lens vesicle

sensory layer of optic cup

pigmented layer of optic cup

external carotid artery

amnion

diencephalon

Figure 8.17

10-mm pig embryo, transverse section through the thyroid rudiment (40X).

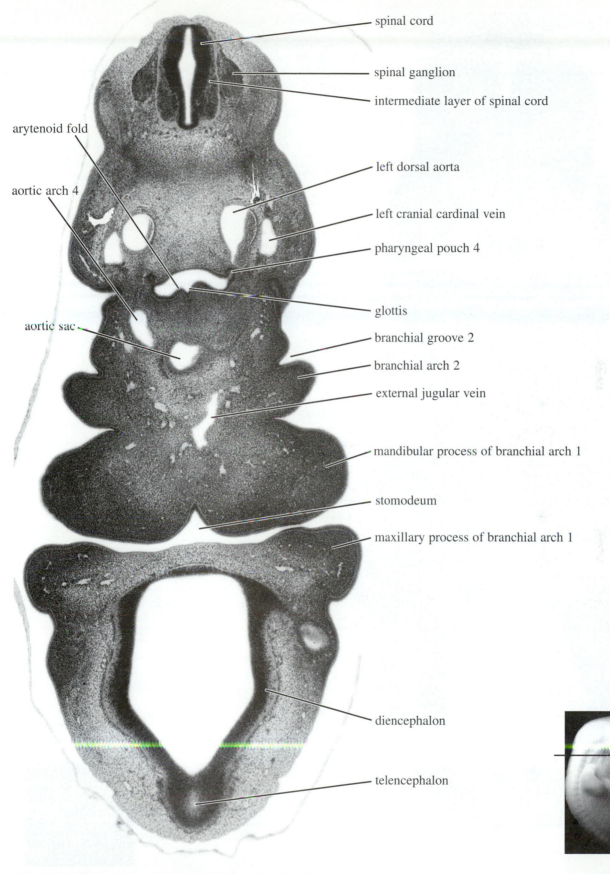

spinal cord

spinal ganglion

intermediate layer of spinal cord

arytenoid fold

aortic arch 4

left dorsal aorta

left cranial cardinal vein

pharyngeal pouch 4

glottis

aortic sac

branchial groove 2

branchial arch 2

external jugular vein

mandibular process of branchial arch 1

stomodeum

maxillary process of branchial arch 1

diencephalon

telencephalon

Figure 8.18

10-mm pig embryo, transverse section through aortic arch 4 (40X).

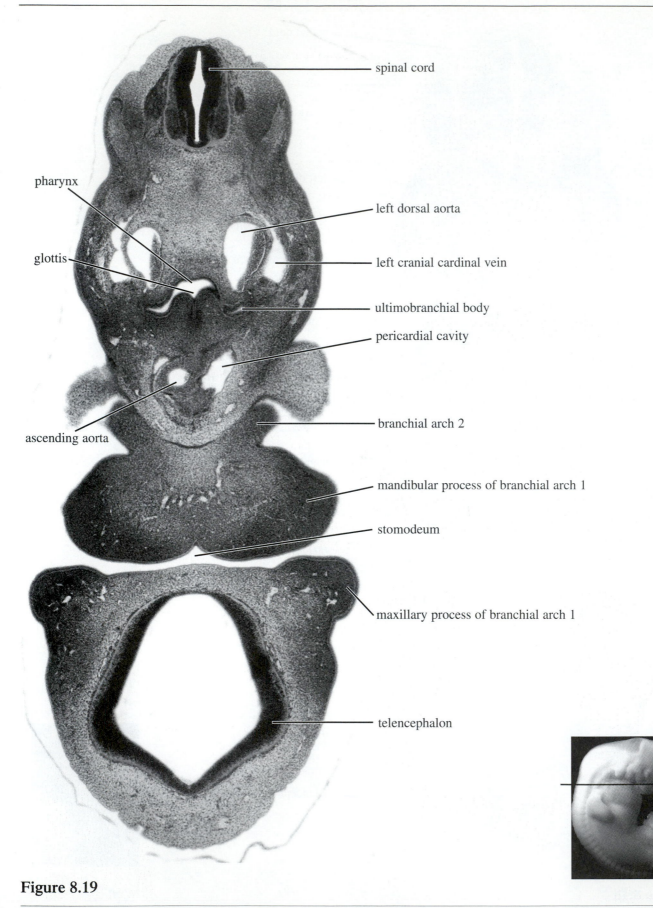

spinal cord

pharynx

left dorsal aorta

glottis

left cranial cardinal vein

ultimobranchial body

pericardial cavity

branchial arch 2

ascending aorta

mandibular process of branchial arch 1

stomodeum

maxillary process of branchial arch 1

telencephalon

Figure 8.19

10-mm pig embryo, transverse section through the ultimobranchial body (40X).

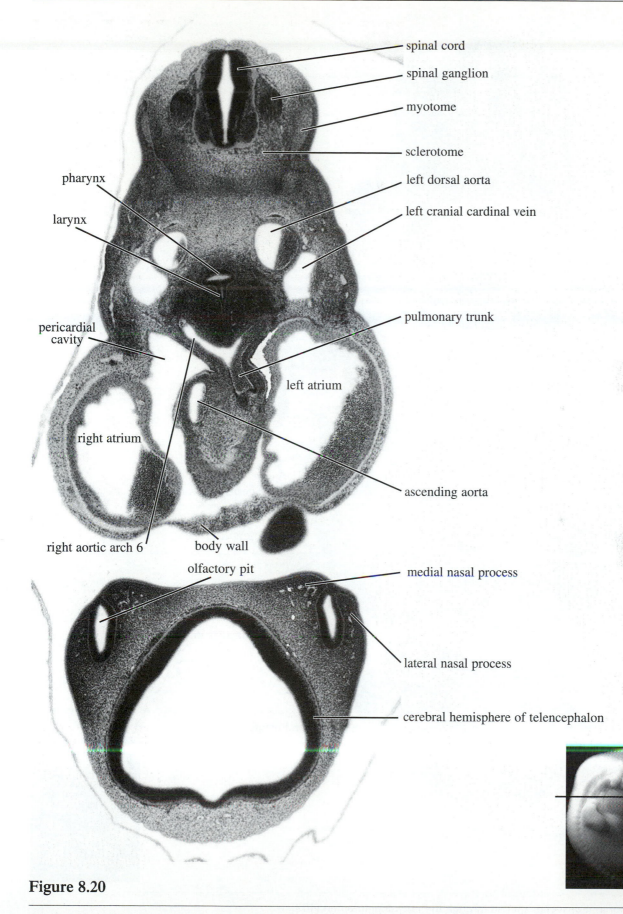

spinal cord

spinal ganglion

myotome

sclerotome

pharynx

larynx

left dorsal aorta

left cranial cardinal vein

pulmonary trunk

pericardial cavity

left atrium

right atrium

ascending aorta

right aortic arch 6

body wall

olfactory pit

medial nasal process

lateral nasal process

cerebral hemisphere of telencephalon

Figure 8.20

10-mm pig embryo, transverse section through the pulmonary trunk (40X).

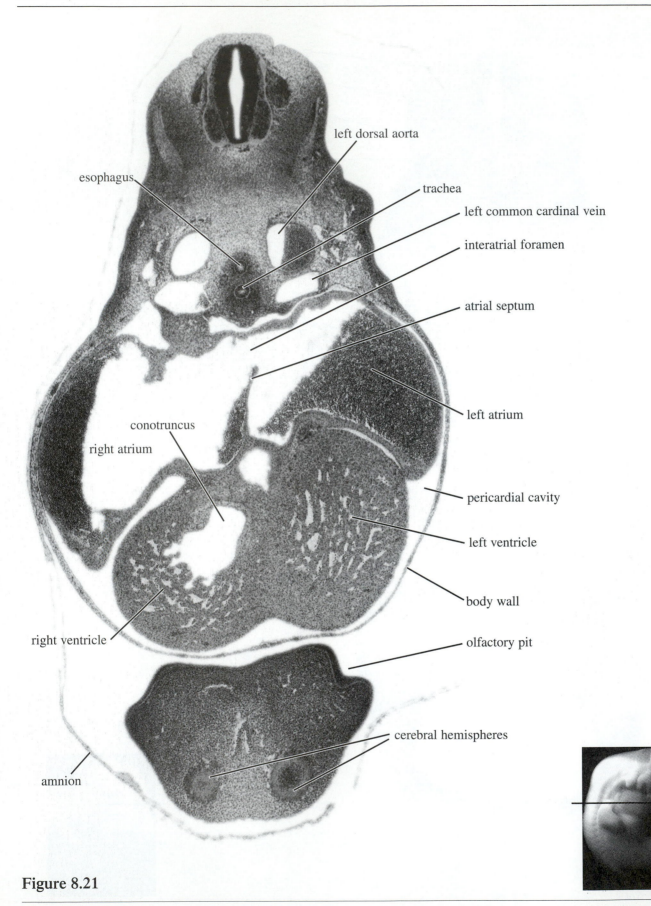

left dorsal aorta

esophagus

trachea

left common cardinal vein

interatrial foramen

atrial septum

conotruncus

right atrium

left atrium

pericardial cavity

left ventricle

body wall

right ventricle

olfactory pit

amnion

cerebral hemispheres

Figure 8.21

10-mm pig embryo, transverse section through the interatrial foramen (40X).

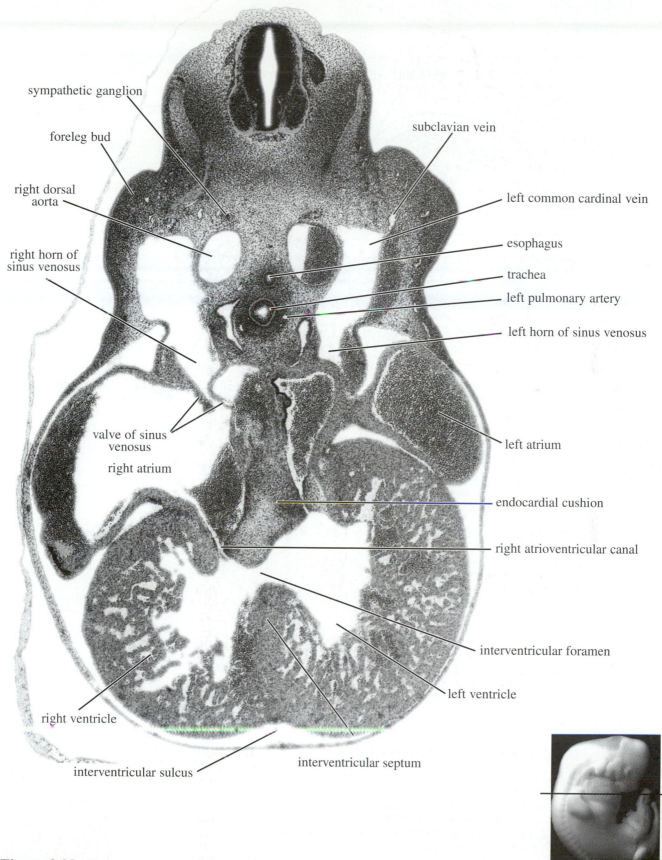

sympathetic ganglion

foreleg bud

right dorsal
aorta

right horn of
sinus venosus

valve of sinus
venosus

right atrium

subclavian vein

left common cardinal vein

esophagus

trachea

left pulmonary artery

left horn of sinus venosus

left atrium

endocardial cushion

right atrioventricular canal

interventricular foramen

left ventricle

right ventricle

interventricular sulcus

interventricular septum

Figure 8.22

10-mm pig embryo, transverse section through the interventricular foramen (40X).

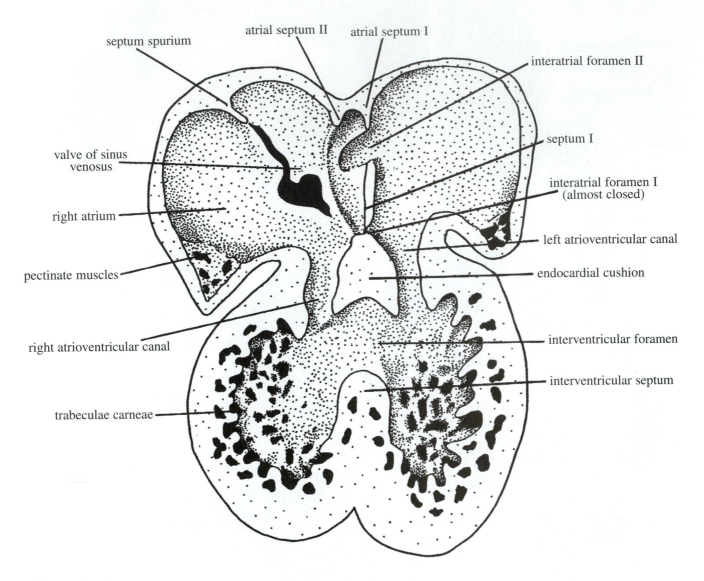

Figure 8.23

Reconstruction of the heart of a 9.4-mm pig embryo; dorsal half of heart, interior view; frontal section.

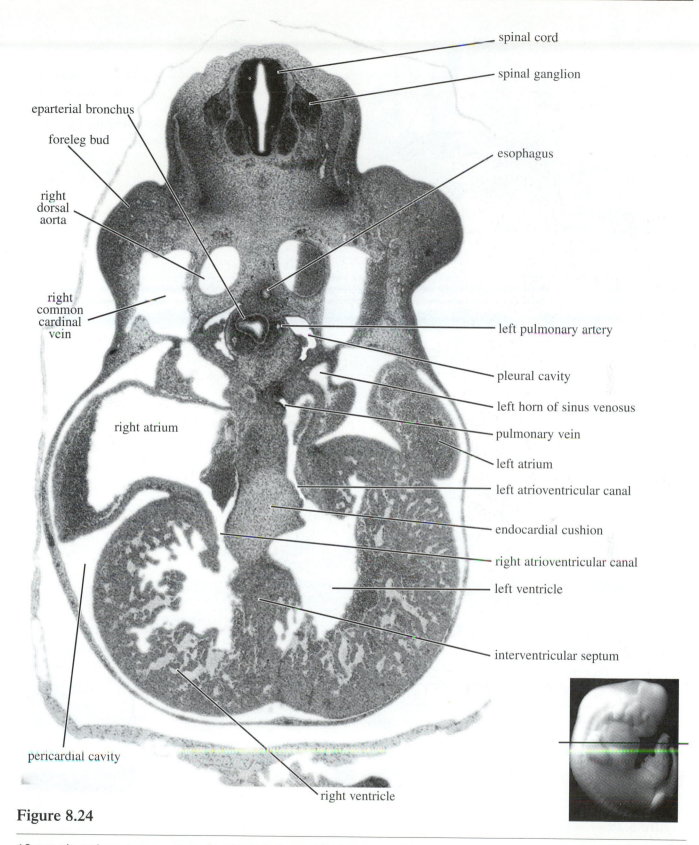

Figure 8.24

10-mm pig embryo, transverse section through the epartial bronchus (40X).

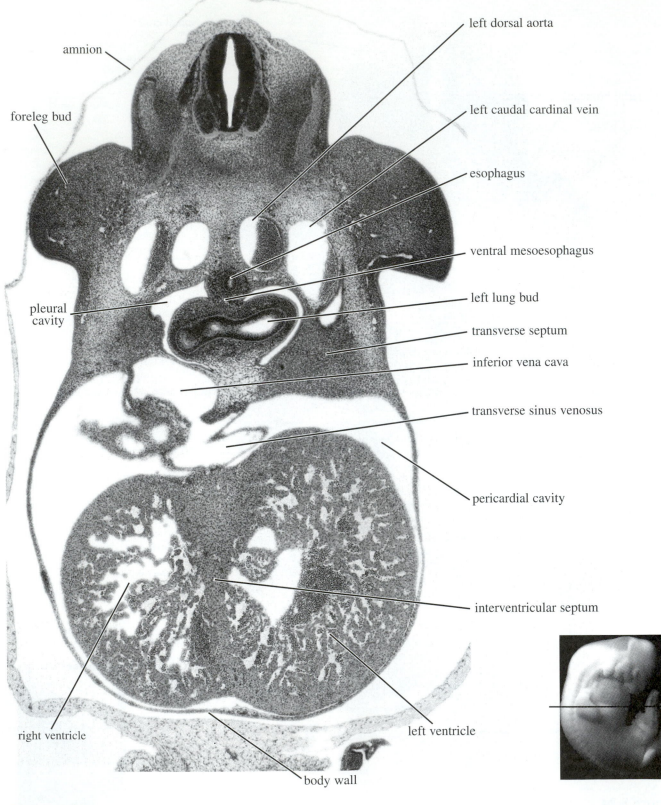

amnion

foreleg bud

left dorsal aorta

left caudal cardinal vein

esophagus

ventral mesoesophagus

pleural cavity

left lung bud

transverse septum

inferior vena cava

transverse sinus venosus

pericardial cavity

interventricular septum

right ventricle

left ventricle

body wall

Figure 8.25

10-mm pig embryo, transverse section through the lung buds (40X).

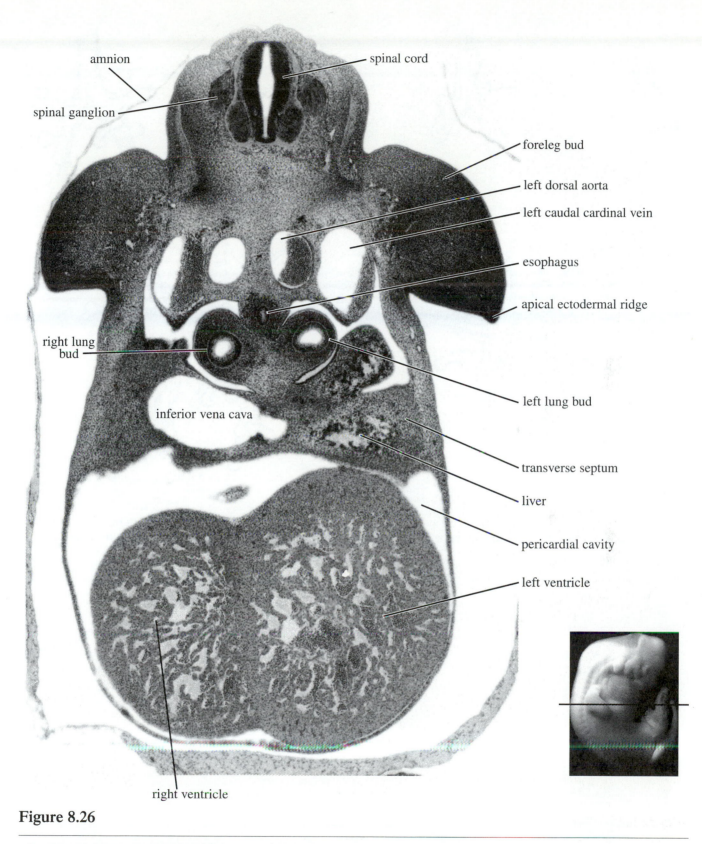

Figure 8.26

10-mm pig embryo, transverse section just caudal to that shown in Figure 256 (40X).

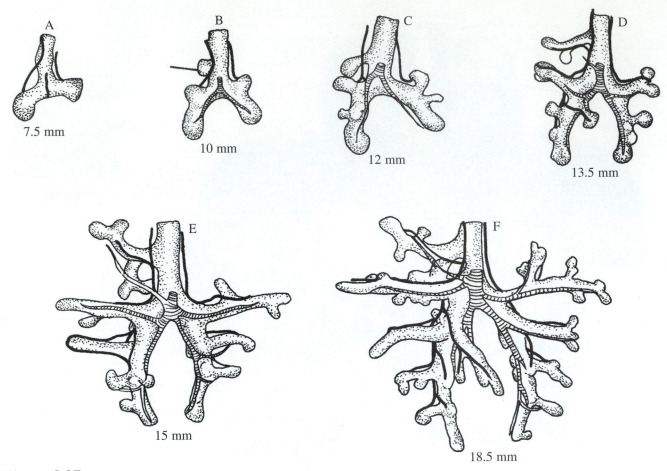

Figure 8.27

Drawings showing the development of the trachea, bronchi and lungs in the pig embryo. Pulmonary arteries are black; pulmonary veins are cross hatched. ep, eparterial bronchus. Compare transverse sections to Figure B (10-mm stage).

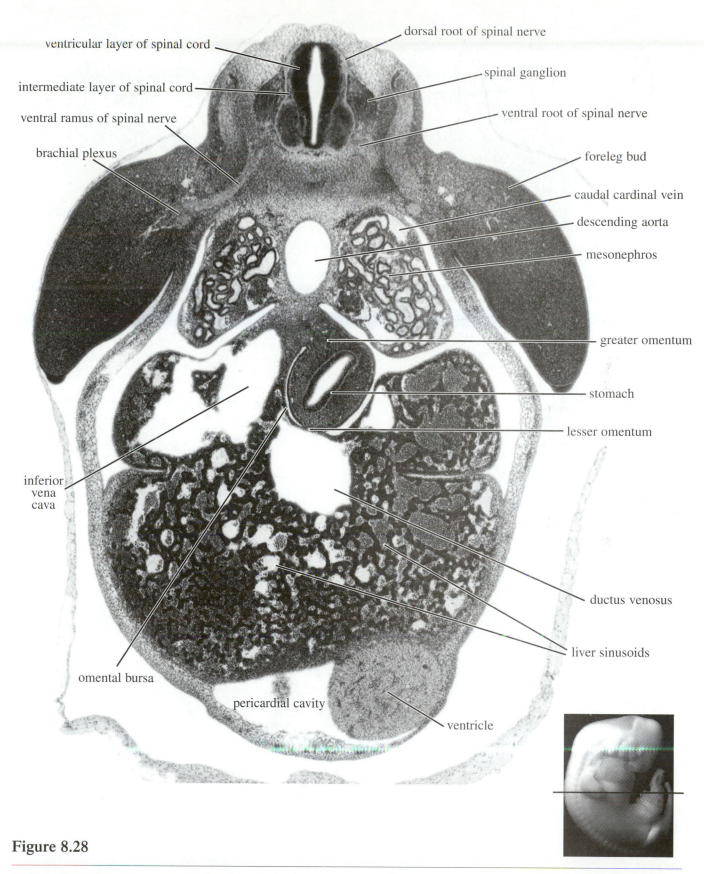

ventricular layer of spinal cord

dorsal root of spinal nerve

intermediate layer of spinal cord

spinal ganglion

ventral ramus of spinal nerve

ventral root of spinal nerve

brachial plexus

foreleg bud

caudal cardinal vein

descending aorta

mesonephros

greater omentum

stomach

lesser omentum

inferior
vena
cava

ductus venosus

liver sinusoids

omental bursa

pericardial cavity

ventricle

Figure 8.28

10-mm pig embryo, transverse section through the ductus venosus (40X).

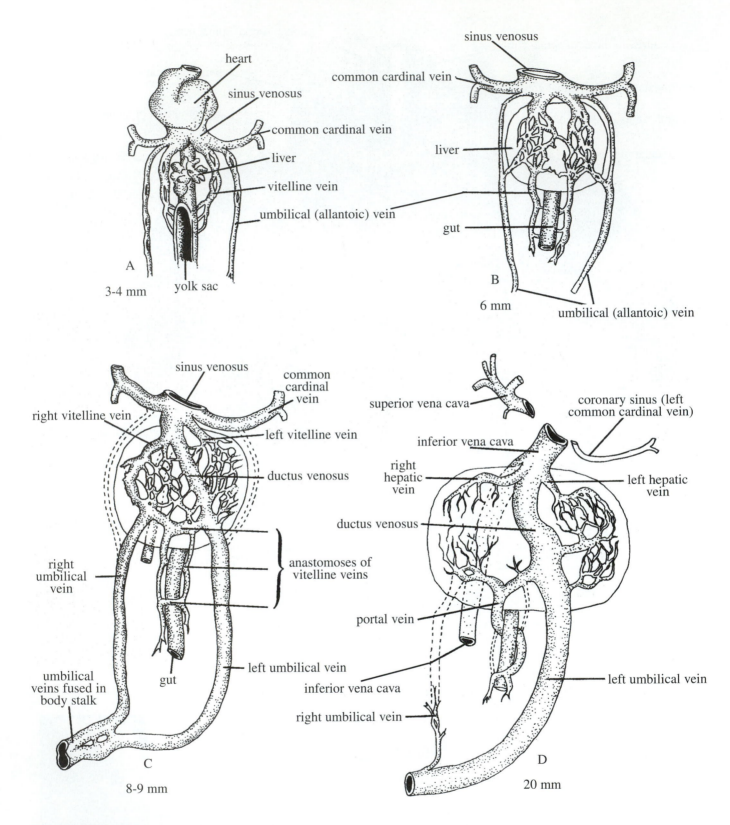

Figure 8.29

Drawings showing the development of the portal circulation and the umbilical veins in the pig embryo. Compare transverse sections to Figure C (8/9-mm stage).

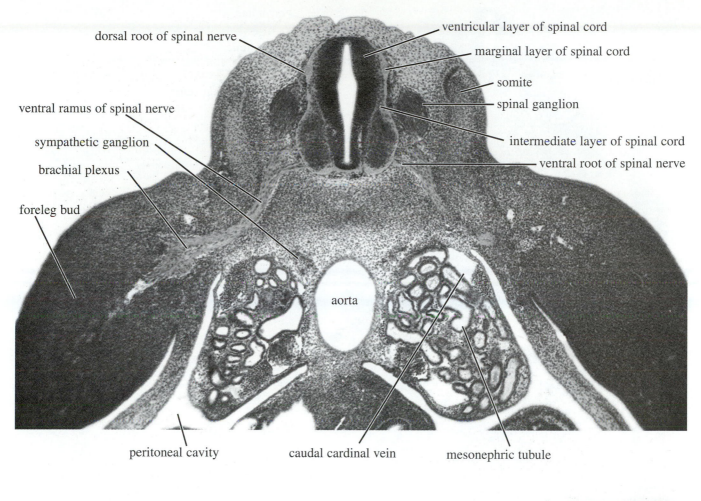

dorsal root of spinal nerve

ventricular layer of spinal cord

marginal layer of spinal cord

somite

ventral ramus of spinal nerve

spinal ganglion

sympathetic ganglion

intermediate layer of spinal cord

brachial plexus

ventral root of spinal nerve

foreleg bud

aorta

peritoneal cavity

caudal cardinal vein

mesonephric tubule

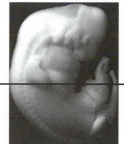

Figure 8.30

10-mm pig embryo, dorsal half of a transverse section through the brachial plexus (75X).

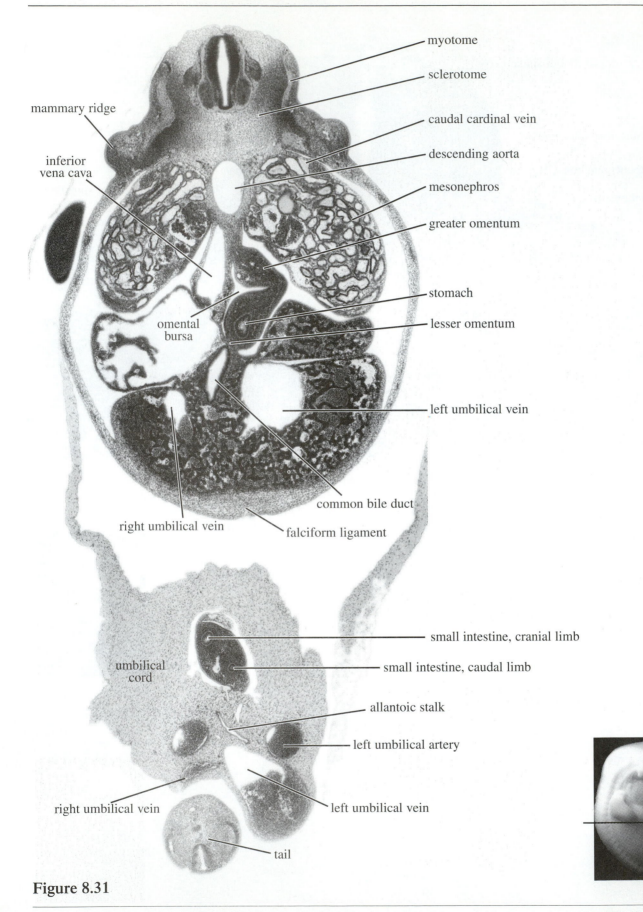

myotome

sclerotome

caudal cardinal vein

descending aorta

mesonephros

greater omentum

mammary ridge

inferior
vena cava

stomach

lesser omentum

omental
bursa

left umbilical vein

right umbilical vein

common bile duct

falciform ligament

umbilical
cord

small intestine, cranial limb

small intestine, caudal limb

allantoic stalk

left umbilical artery

right umbilical vein

left umbilical vein

tail

Figure 8.31

10-mm pig embryo, transverse section through the common bile duct (40X).

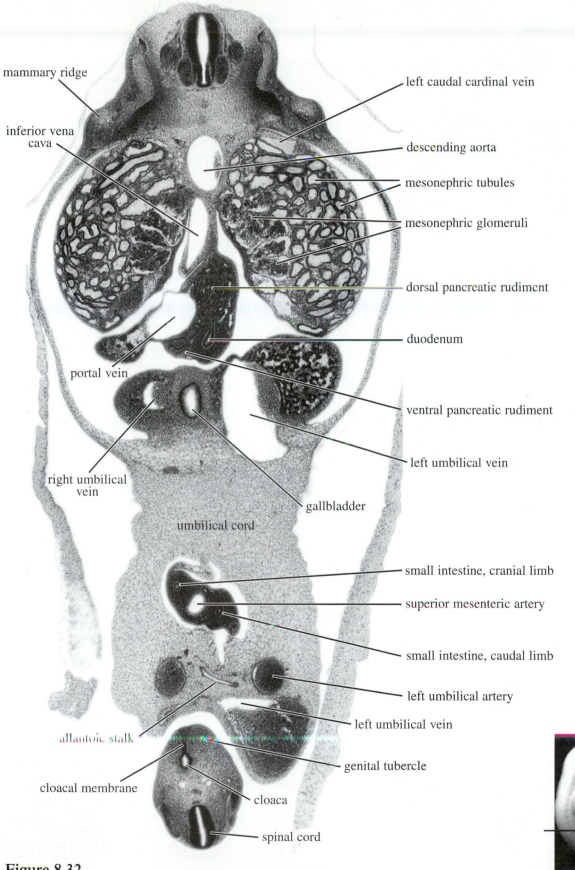

mammary ridge

inferior vena
cava

portal vein

right umbilical
vein

umbilical cord

allantoic stalk

cloacal membrane

Figure 8.32

left caudal cardinal vein

descending aorta

mesonephric tubules

mesonephric glomeruli

dorsal pancreatic rudiment

duodenum

ventral pancreatic rudiment

left umbilical vein

gallblatter

small intestine, cranial limb

superior mesenteric artery

small intestine, caudal limb

left umbilical artery

left umbilical vein

genital tubercle

cloaca

spinal cord

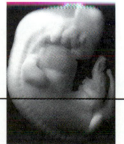

10-mm pig embryo, transverse section through the gallbladder (40X).

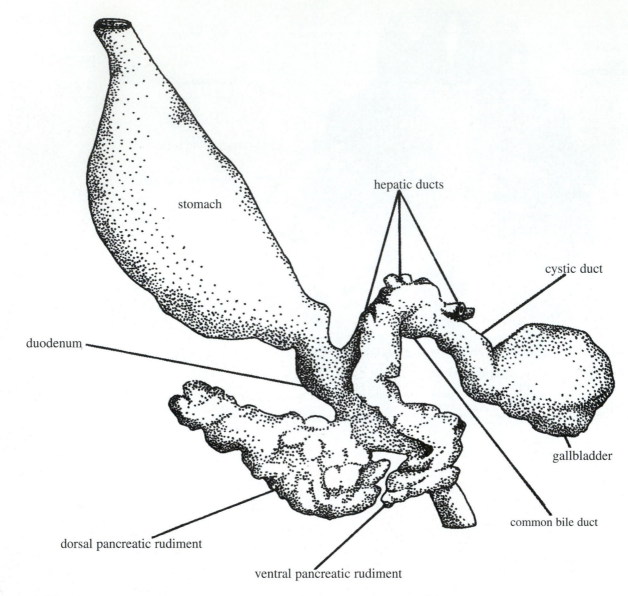

stomach

hepatic ducts

cystic duct

duodenum

gallbladder

common bile duct

dorsal pancreatic rudiment

ventral pancreatic rudiment

Figure 8.33

Reconstruction of the stomach, duodenum and gallbladder of a 9.4-mm pig embryo showing the rudiments of the pancreas.

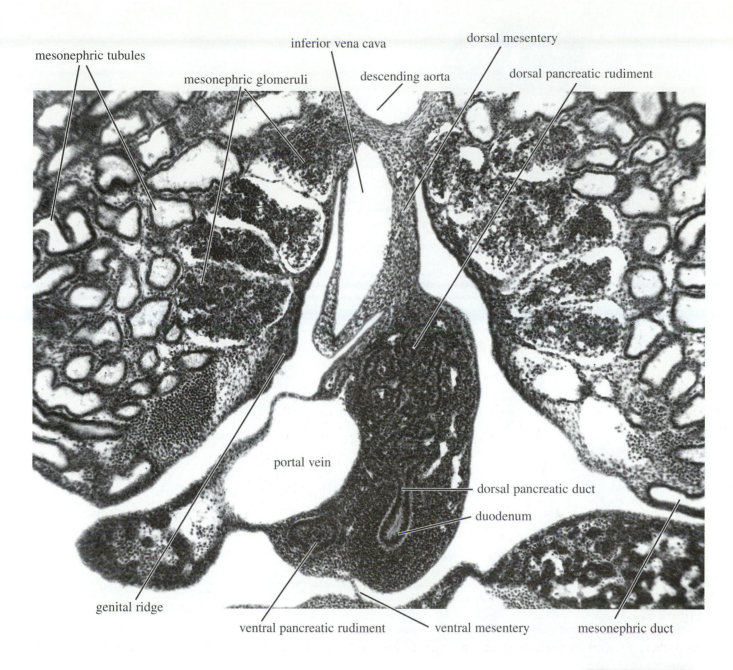

mesonephric tubules

mesonephric glomeruli

inferior vena cava

descending aorta

dorsal mesentery

dorsal pancreatic rudiment

portal vein

dorsal pancreatic duct

duodenum

genital ridge

ventral pancreatic rudiment

ventral mesentery

mesonephric duct

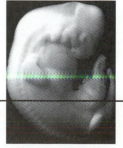

Figure 8.34

10-mm pig embryo, middle part of a transverse section through the pancreatic rudiments (95X).

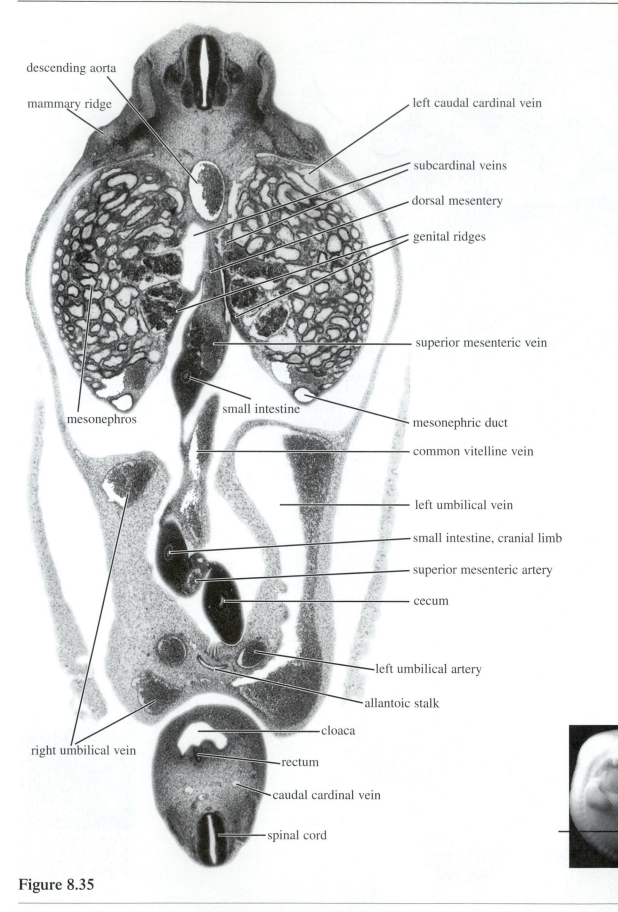

descending aorta

mammary ridge

left caudal cardinal vein

subcardinal veins

dorsal mesentery

genital ridges

superior mesenteric vein

mesonephros

small intestine

mesonephric duct

common vitelline vein

left umbilical vein

small intestine, cranial limb

superior mesenteric artery

cecum

left umbilical artery

allantoic stalk

right umbilical vein

cloaca

rectum

caudal cardinal vein

spinal cord

Figure 8.35

10-mm pig embryo, transverse section through the genital ridges (40X).

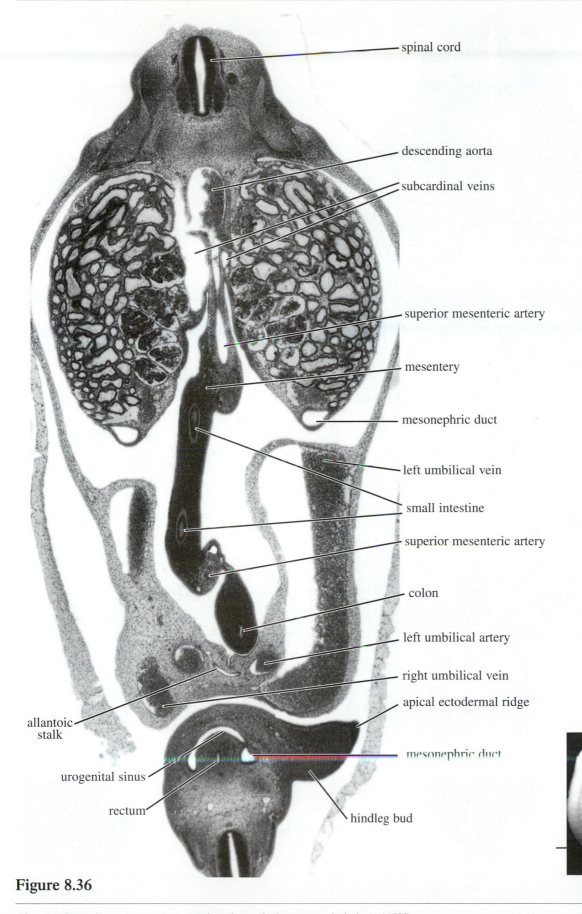

spinal cord

descending aorta

subcardinal veins

superior mesenteric artery

mesentery

mesonephric duct

left umbilical vein

small intestine

superior mesenteric artery

colon

left umbilical artery

right umbilical vein

apical ectodermal ridge

mesonephric duct

hindleg bud

allantoic stalk

urogenital sinus

rectum

Figure 8.36

10-mm pig embryo, transverse section through the urogenital sinus (40X).

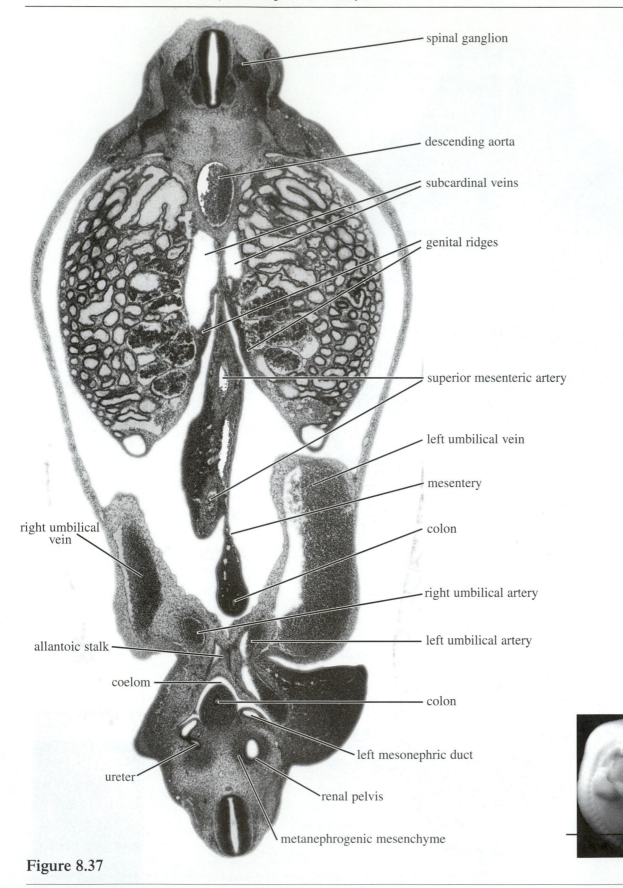

Figure 8.37

10-mm pig embryo, transverse section through the metanephros (40X).

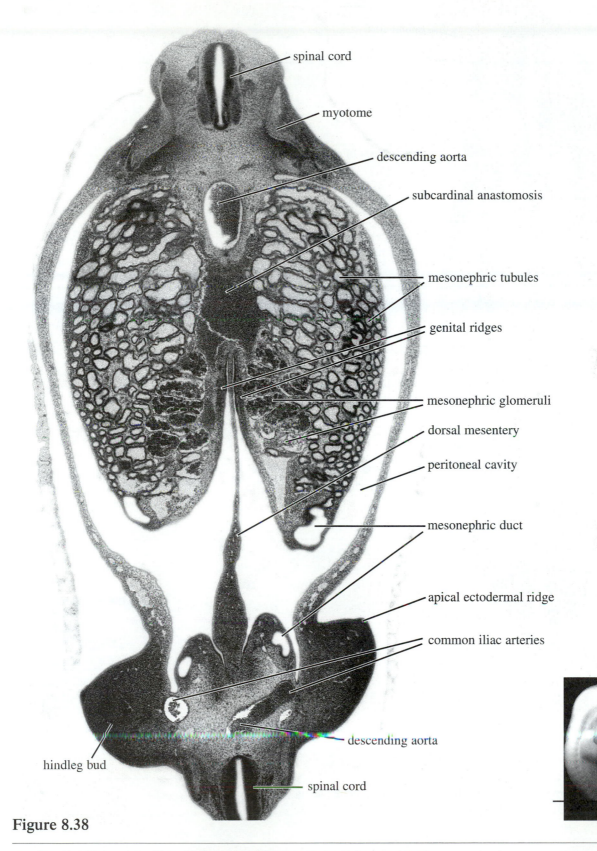

spinal cord

myotome

descending aorta

subcardinal anastomosis

mesonephric tubules

genital ridges

mesonephric glomeruli

dorsal mesentery

peritoneal cavity

mesonephric duct

apical ectodermal ridge

common iliac arteries

descending aorta

hindleg bud

spinal cord

Figure 8.38

10-mm pig embryo, transverse section through the common iliac artery (40X).

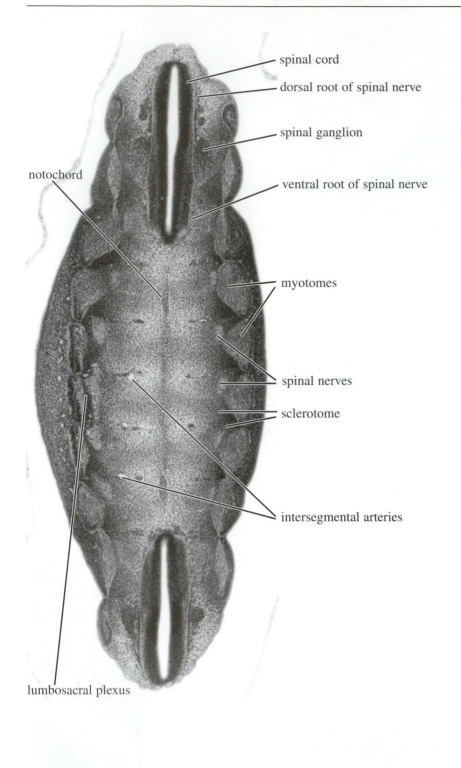

spinal cord

dorsal root of spinal nerve

spinal ganglion

notochord

ventral root of spinal nerve

myotomes

spinal nerves

sclerotome

intersegmental arteries

lumbosacral plexus

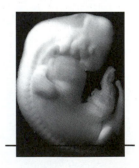

Figure 8.39

10-mm pig embryo, transverse section through the lumbosacral plexus (40X).

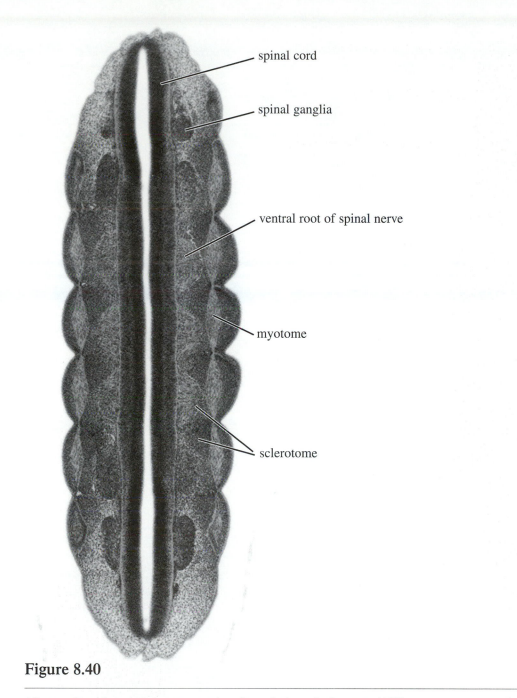

spinal cord

spinal ganglia

ventral root of spinal nerve

myotome

sclerotome

Figure 8.40

10-mm pig embryo, transverse section through the spinal nerves (40X).

Human Reproduction

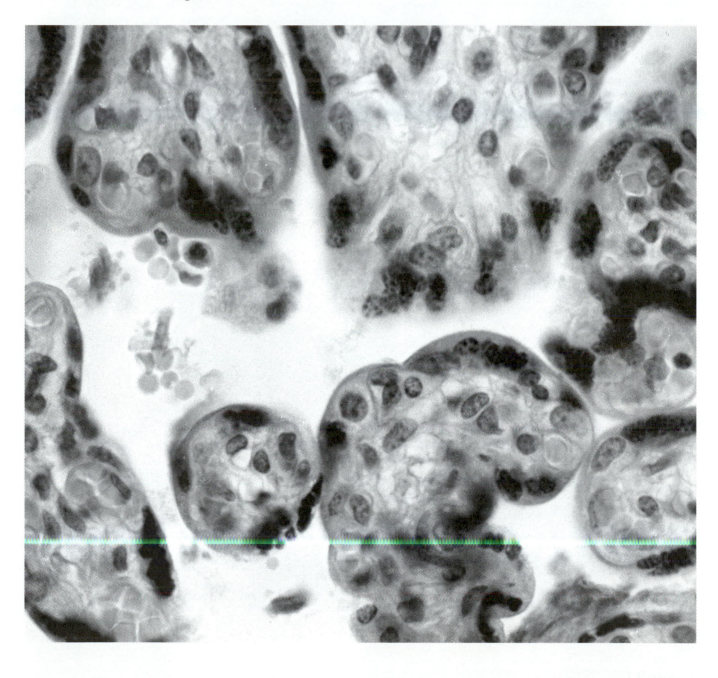

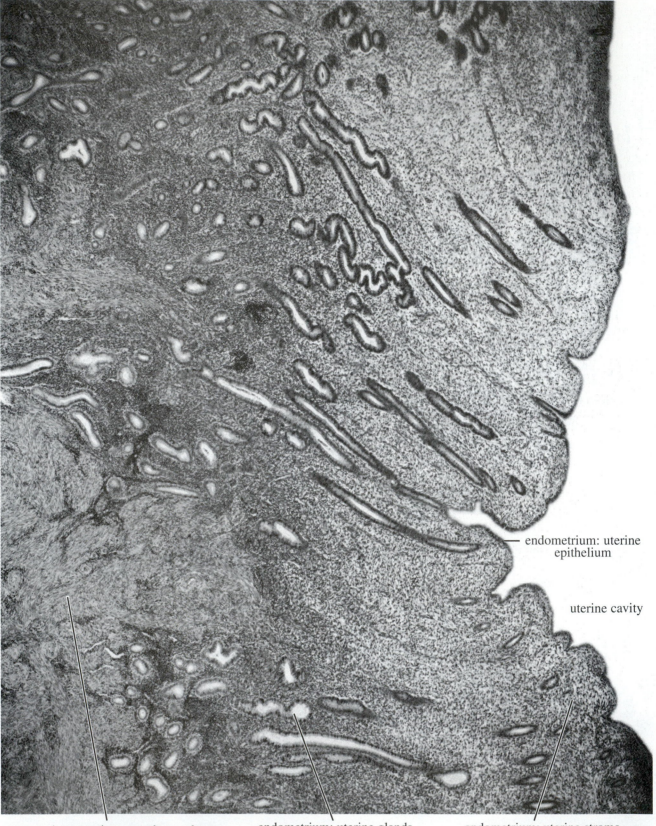

endometrium: uterine
epithelium

uterine cavity

myometrium: uterine smooth muscle   endometrium: uterine glands   endometrium: uterine stroma

Figure 9.1

Human uterus, section, second week of menstrual cycle (70X).

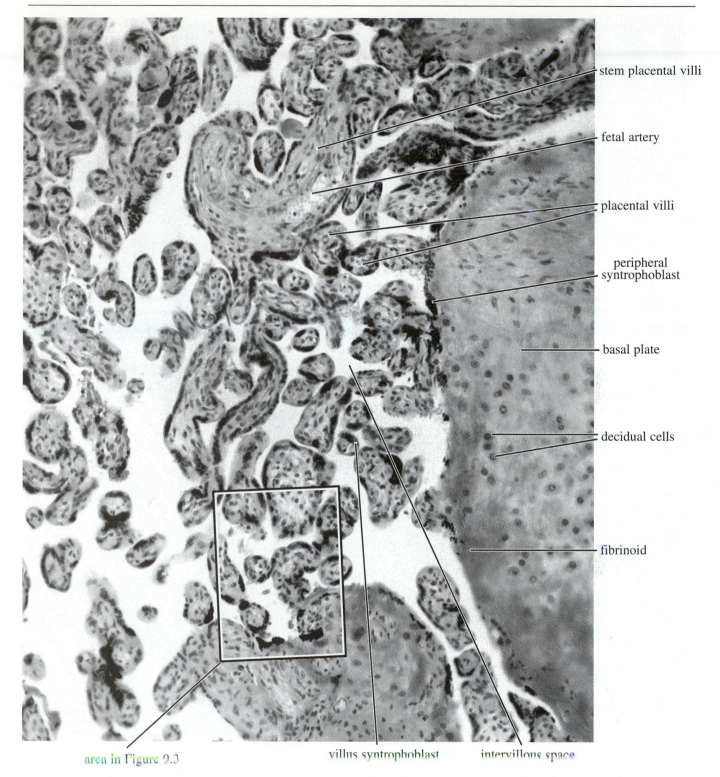

stem placental villi

fetal artery

placental villi

peripheral
syntrophoblast

basal plate

decidual cells

fibrinoid

area in Figure 9.3

villus syntrophoblast

intervillous space

Figure 9.2

Human placenta, section (200X).

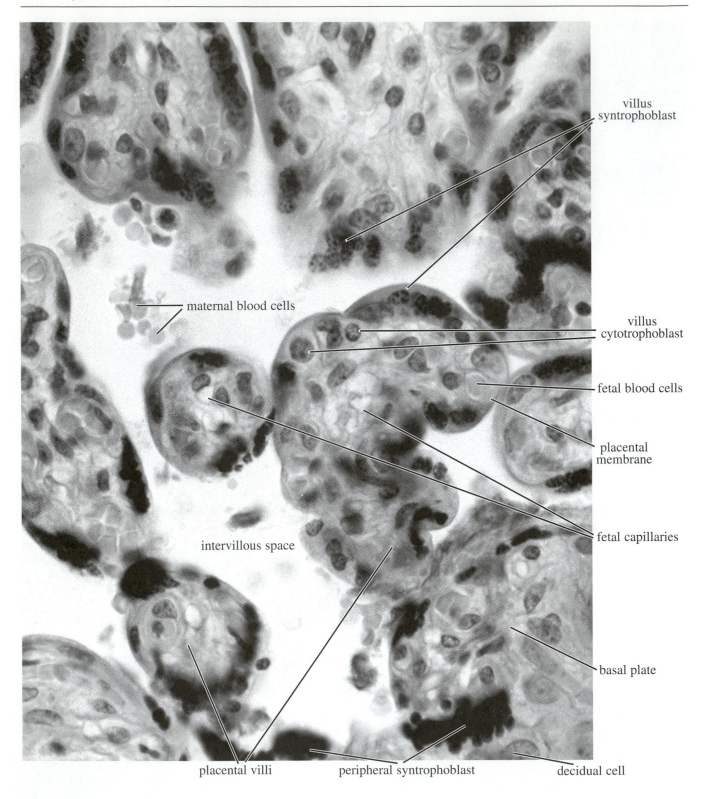

villus
syntrophoblast

maternal blood cells

villus
cytotrophoblast

fetal blood cells

placental
membrane

fetal capillaries

intervillous space

basal plate

placental villi

peripheral syntrophoblast

decidual cell

Figure 9.3

Human placenta, section (940X).

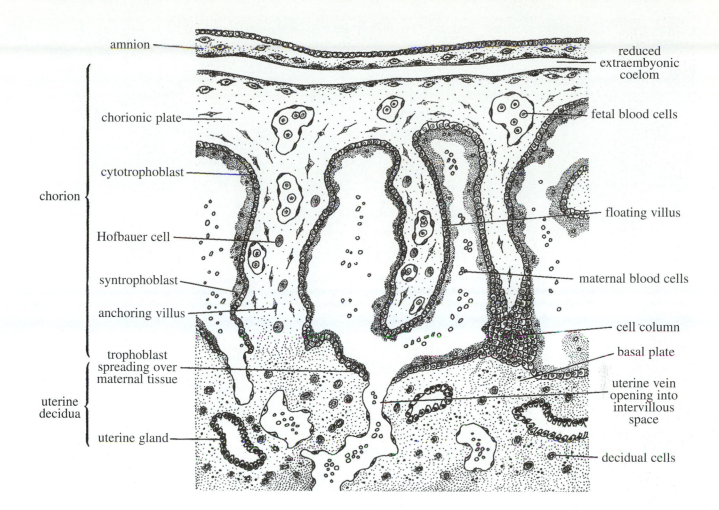

amnion

reduced
extraembyonic
coelom

chorionic plate

fetal blood cells

cytotrophoblast

chorion

floating villus

Hofbauer cell

syntrophoblast

maternal blood cells

anchoring villus

cell column

trophoblast
spreading over
maternal tissue

basal plate

uterine
decidua

uterine vein
opening into
intervillous
space

uterine gland

decidual cells

Figure 9.4

Semischematic drawing showing the relations of the chorionic villi and trophoblast to maternal endometrial tissues of the human placenta.

Credits

Figures 1.3, 5.2, 5.19, 6.16, 6.45, 6.57, 7.3, 8.6

Adapted from O.E. Nelson, Comparative Embryology of the Vertebrates. Copyright 1953 by the Blaskiston Company. Used with the permission of the McGraw-Hill Book Company.

Figure 1.6

Adapted from G. Karp and N.J. Berrill, Development, Second Edition. Copyright 1981 by the McGraw-Hill Book Company. Used with the permission of the McGraw-Hill Book Company.

Figures 1.11, 6.13-6.15, 7.67

Adapted from A.F. Huettner, Fundamentals of Comparative Embryology of the Vertebrates. Copyright 1941 by the Macmillan Publishing Company. Used with the permission of the Macmillan Publishing Company.

Figures 2.7, 2.8, 4.2, 4.3

Adapted from E.B. Wilson, The Cell in Development and Inheritance. Copyright 1925 by the Macmillan Publishing Company. Used with the permission of the Macmillan Publishing Company.

Figure 3.1

Adapted from J.E. Sulston and H.R. Horvitz, Post-embryonic cell lineages of the nematode, *Caenorhabditis elegans. Developmental Biology* 56:111. Copyright 1977 by Academic Press. Used with the permission of Academic Press.

Figure 3.2

Adapted from J.E. Sulston, E. Schierenberg, J.G. White and J.N. Thomson, The embryonic cell lineage of the nematode *Caenorhabditis elegans. Developmental Biology* 100:90. Copyright 1983 by Academic Press. Used with the permission of Academic Press.

Figures 4.19, 7.12, 7.13

Adapted from B.I. Balinsky assisted by B.C. Babian, An Introduction to Embryology, Fifth Edition. Copyright 1981 by the W.B. Saunders Company. Used with the permission of the W.B. Saunders Company, a division of CBS Publishing.

Figures 5.3, 5.4

Adapted from E.G. Conklin, The embryology of *Amphioxus. Journal of Morphology* 54:69. Copyright 1932 by the Wistar Institute Press. Used with the permission of John Wiley and Sons, Inc.

Figure 5.20

Adapted from R.S. McEwen, Vertebrate Embryology. Copyright 1923 by Henry Holt and Company. Used with the permission of Henry Holt and Company.

Figures 6.3, 6.5-6.7, 7.6

Adapted from E. Witschi, Development of Vertebrates. Copyright 1956 by the W.B. Saunders Company. Used with the permission of the W.B. Saunders Company, a division of CBS College Publishing.

Figure 6.17

Adapted from J. Pasteels, New observations concerning maps of presumptive areas of the young amphibian gastrula (*Amblystoma* and *Discoglossus*). *Journal of Experimental Zoology* 89:225. Copyright 1943 by the Wistar Institute Press. Used with the permission of John Wiley and Sons, Inc.

Figures 7.5, 7.7

Adapted from H.L. Wieman, An Introduction to Vertebrate Embryology. Copyright 1949 by the McGraw-Hill Book Company. Used with the permission of the McGraw-Hill Book Company.

Figures 7.21, 7.22, 7.28, 7.29, 7.32, 7.33, 7.45, 7.48, 7.85, 7.89

Adapted from B.M. Patten, Early Embryology of the Chick, Fifth Edition. Copyright 1951 by the McGraw-Hill Book Company. Used with the permission of the McGraw-Hill Book Company.

Figures 8.2, 8.7, 8.23, 8.27, 8.29, 8.33

Adapted from B.M. Patten, Embryology of the Pig, Third Edition. Copyright renewed 1959 by B.M. Patten. Used with the permission of the McGraw-Hill Book Company.

Figure 8.16

Adapted from L.B. Arey, Developmental Anatomy, Seventh Edition Revised. Copyright 1974 by the W.B. Saunders Company. Used with the permission of the W.B. Saunders Company.

Figure 9.4

Adapted from B.M. Patten, Human Embryology. Copyright 1968 by the McGraw-Hill Book Company. Used with the permission of the McGraw-Hill Book Company.

Glossary, Synopsis of Development and Index

AB blastomeres (including ABa, ABp, right and left)
see founder cells

accessory cells
Figure 4.1
Ovarian nurse cells of sea urchins containing yolk granules, lipid droplets and glycogen; these contents are probably transferred to the oocytes enveloped by the nurse cells.

accessory cleavage
Figure 7.7
Small cleavage furrows in the blastoderm of birds that form around supernumerary sperm nuclei; they subsequently degenerate.

accessory ganglia
Figure 8.6
A chain of small ganglia along the myelencephalon that contributes sensory fibers to cranial nerves X and XII.

adenohypophysis
Figures 6.17, 6.21, 6.26, 6.29, 6.34, 6.41, 6.43, 6.45, 6.57
also see hypophysis
The anterior pituitary gland.

adhesive glands
Figures 6.17, 6.21-6.23, 6.28, 6.29, 6.34-6.36, 6.40, 6.48, 6.49
Paired ectodermal thickenings on the ventral side of the head of anuran tadpoles; they secrete adhesive mucus which is used for attachment.

air chamber
Figures 7.6, 7.48
An air-filled space between the inner and outer shell membranes of bird eggs; formed by the shrinkage of egg contents caused by water loss through the porous shell; provides a "breathing space" for the chick before hatching.

albumen
Figures 7.5, 7.6, 7.48
also see chalaza
The "white" of bird eggs; provides protein and water for the growing embryo; consists of fluid and dense components including the chalaza; composed of the proteins ovalbumen conalbumen, ovomucin, avidin and lysozyme; the latter two substances protect the embryo from microorganisms; produced by the magnum segment of the hen's oviduct under the stimulus of estrogen. *Synonym:* egg white.

albumen sac
Figure 7.48
also see chorioallantois
An extension of the chorioallantois around the egg albumen in bird embryos; absorbs the albumen.

allantoic arteries
Figures 7.85, 7.89
also see umbilical arteries
The arterial supply of the allantois; in mammals, they constitute the umbilical arteries supplying the fetal component of the placenta; they arise as paired vessels from the caudal portions of the dorsal aortae; they cease to function at birth and degenerate. *Synonym:* umbilical arteries.

allantoic cavity
Figure 7.48
also see allantois
The cavity of the allantois, lined by endoderm of the splanchnopleure, communicates with the cloaca via the allantoic stalk; in bird embryos and in some mammals, it stores the embryo's urine.

allantoic stalk
Figures 7.48, 7.88, 8.2, 8.6, 8.9, 8.31, 8.32, 8.35-8.37
A canal in the umbilical cord connecting the cloaca or, later, the urogenital sinus with the allantois; carries urine to the allantois in many amniotes.

allantoic veins
Figures 7.67, 7.76, 7.79, 7.88, 8.29
see umbilical veins

allantois
Figures 7.48, 7.62, 7.63, 7.68, 7.70, 7.82, 7.84, 7.86, 7.88
An extraembryonic membrane of amniotes; grows out of the hindgut to fuse with the chorion and to form the chorioallantois; vascularized by the umbilical vessels; functions as the main embryonic respiratory organ of birds, stores embryonic urine and forms the albumen sac; in mammals, it contributes to the placenta.

amniocardiac vesicles
Figures 7.20, 7.24
also see coelom; extraembryonic coelom; intraembryonic coelom; pericardial cavity
Paired dilations of the coelom in the heart region (pericardial cavity) and extending laterally into the extraembryonic area where they participate in formation of the amnion.

amnion
Figures 7.47-7.49, 7.51, 7.52, 7.55-7.57, 7.59, 7.60, 7.68-7.71, 7.75-7.81, 7.83, 7.87, 7.88, 8.2, 8.10, 8.11, 8.17, 8.25, 8.26, 9.4
also see amniotic folds
An extraembryonic membrane of amniotes that encloses the embryo and amniotic fluid; usually arises from folds of somatopleure.

amniotic cavity
Figures 7.47, 7.48, 7.50, 7.68, 7.71-7.75, 7.80, 7.82, 7.86, 7.89
also see amnion
The lumen of the amnion; contains amniotic fluid and the embryo.

amniotic folds
Figures 7.34-7.36, 7.43, 7.44, 7.48, 7.60-7.62, 7.82
also see amnion; chorion
Folds of somatopleure surrounding the embryo, arising first cranial to the head, then extending along the sides and encircling the tail; the folds are drawn over the embryo, enclosing it with two membranes, an inner amnion and an outer chorion.

anal arms (lobes)
Figures 4.15-4.19, 4.32
also see oral arms
Slender, paired extensions of the ventral body wall of the pluteus larva; they are supported by skeletal rods and bear ciliated bands; they function to stabilize and propel the larva.

anal pit
see proctodeum

anaphase
Figures 1.3, 5.2
also see metaphase; prophase; telophase
A phase in cell division in which chromosomes are moving along spindle fibers toward the spindle poles (centrioles).

anastomosis of basilar artery and internal carotid artery
Figure 8.12
A ring-like linkage of basilar and internal carotid arteries encircling the infundibulum; forms the circle of Willis.

anastomoses of vitelline veins
Figure 8.29
After the anastomosis, certain segments of vitelline veins atrophy, leaving a single channel, the portal vein, which is therefore derived from parts of both right and left vitelline veins.

anchoring villi
Figure 9.4
also see floating villi; placental villi; stem placental villi
Placental villi attached to the uterine decidua.

animal cap
Figure 6.11
The roof of the blastocoel in the amphibian blastula.

animal hemisphere
Figures 4.9, 5.13
also see animal pole; vegetal hemisphere; micromeres
That half of the egg or early embryo containing the least yolk and consequently the most active cytoplasm. The animal pole lies at its center and opposite the vegetal pole.

animal plate
Figures 4.12-4.13
A thickening of ectoderm near the animal pole of sea urchin embryos; bears long cilia of the apical tuft.

animal pole
Figures 4.10, 4.11, 4.26-4.29, 5.12, 6.10, 6.17
The end of the embryonic axis centered in the most active region; opposite the vegetal pole.

anterior
Figures 3.12, 3.14, 3.17-3.22
The head end of the embryo.

antrum
see follicular cavity

anus
Figures 3.1, 4.15, 4.19, 4.31-5.1, 5.4, 5.18, 6.40, 6.41, 6.45, 6.55, 6.57
The caudal opening of the digestive tract; derives from the blastopore in the sea urchin and starfish and from the proctodeum in *Amphioxus* and vertebrates.

aorta
see descending aorta

aortic arches
Figure 6.49
also see aortic arches 1, 2, 3, 4, 5, 6
Paired arterial connections between the dorsal and ventral aorta (or later, aortic sac) and encircling the pharynx; they lie within the branchial arches and arise from the head mesenchyme; some degenerate, whereas others persist as adult arteries.

aortic arch 1
Figures 6.45, 7.32, 7.33, 7.35, 7.45, 7.46, 7.52, 7.53, 7.66, 8.6
The cranial member of a series of six paired arterial connections between the dorsal and ventral aortae; they lie within branchial arch 1 and degenerate at an early stage.

aortic arch 2
Figures 7.45, 7.46, 7.54, 7.66, 7.69, 7.70, 7.72, 7.73, 7.85, 7.88, 8.6
The second of a series of six paired arterial connections between the dorsal and ventral aortae; they lie within the branchial arch 2 and degenerate at an early stage.

aortic arch 3
Figures 6.31, 6.45, 7.45, 7.46, 7.64, 7.66, 7.69, 7.70, 7.72-7.74, 7.85, 7.87, 7.88, 8.6, 8.7, 8.9, 8.15, 8.17
The third of a series of six paired arterial connections between the dorsal and ventral aortae; they lie within the branchial arch 3 and persist as part of the internal carotid arteries; in mammals, they also form part of the common carotid arteries.

aortic arch 4
Figures 6.31, 7.66, 7.85, 7.87, 7.88, 8.2, 8.6, 8.7, 8.18
The fourth of a series of six paired arterial connections between the dorsal and ventral aortae; they lie within branchial arch 4. The right fourth aortic arch forms the arch of the aorta in birds, whereas the left fourth arch forms the arch of the aorta of mammals; in amphibians both the right and left persist as aortae.

aortic arch 5
Figure 8.7
The sixth of a series of paired arterial connections between the dorsal and ventral aortae; vestigial and inconstant; they atrophy completely.

aortic arch 6
Figures 7.85, 8.6, 8.7, 8.20
The sixth of a series of paired arterial connections between the dorsal and ventral aortae; they form part of the pulmonary arteries of tetrapods and, in late embryonic stages, one forms the ductus arteriosus.

aortic sac
Figures 7.45, 7.46, 7.54, 7.66, 7.68, 7.74, 7.86, 8.3, 8.17, 8.18
Formed from fusion of the ventral aortae, ventral to the pharynx. The aortic arches arise from the aortic sac.

apical ectodermal ridge
Figures 7.79, 7.82, 8.26, 8.36, 8.38
An ectodermal thickening on the distal margin of the limb buds of amniotes.

archenteric vesicle
Figure 4.29
also see coelomic sacs
The inner expanded end of the archenteron of the starfish gastrula; forms mesenchyme which scatters through the blastocoel and then forms the coelomic sacs.

archenteron
Figures 4.12-4.14, 4.28-4.30, 5.3, 5.4, 5.14, 5.15, 5.20, 6.11-6.13, 6.15, 6.16
The primitive gut formed by gastrulation movements; at least part of its wall is endoderm; opens through a blastopore to the exterior. *Synonym:* gastrocoel.

archenteron roof
Figure 6.12
The dorsal wall of the archenteron; in amphibians, it forms the notochord and the dorsal mesoderm; in *Amphioxus,* it forms the notochord and somites.

area opaca
Figures 7.8, 7.9, 7.15-7.18
The peripheral zone of the chick blastoderm; it is attached to the underlying yolk; it surrounds the area pellucida.

area opaca vasculosa
Figures 7.14, 7.19, 7.20
also see blood islands
The inner region of the area opaca; it contains thickenings of splanchnic mesoderm, the blood islands, which differentiate into red blood cells and vitelline blood vessels.

area opaca vitellina
Figure 7.19
The outer region of the area opaca; it extends outward from the area vasculosa and is temporarily free of blood and blood vessels.

area pellucida
Figures 7.8, 7.9, 7.12, 7.14, 7.19, 7.20
The relatively transparent central region of the chick blastoderm; it is underlain by the fluid-filled subgerminal cavity. The primitive streak and embryonic axis form within it.

artifact
Figures 6.4, 6.10, 7.53
A change in the structure of a microscopic preparation caused by the processing technique; it often consists of spaces or cracks from shrinkage, or as granules and fibers precipitated by fixing fluids and solvents.

arytenoid folds
Figures 8.16, 8.18
Paired ridges lateral to the glottis, derived from branchial arches 4 and 5; they contribute to the wall of the larynx.

ascending aorta
Figures 8.8, 8.19, 8.20
The part of the aorta that extends cranially from the heart; it forms from the longitudinal division of the conotruncus and from the aortic sac.

atretic follicle
Figures 1.7, 1.8
A degenerating ovarian follicle in mammals. Atresia may appear at any stage of follicular growth, destroying the oocyte, follicle cells and theca; in some species (cat), atresia involves dispersal of some follicle cells, which then form interstitial cells.

atrial septum
Figure 8.21
also see interatrial foramen; atrial septum I; atrial septum II
A partition arising from the wall of the primitive atrium near its sagittal plane in tetrapods; eventually it divides the primitive atrium into right and left atria. An interatrial foramen in the septum prevents complete separation of the atria until the foramen closes after hatching in birds or after birth in mammals.

atrial septum I
Figure 8.23
also see atrial septum II, interatrial foramen
A crescent-shaped median partition of the atrium dividing it into a right and left atrium; the atrial septum I is incomplete; the original opening in it is the interatrial foramen I, which allows blood to pass from the right to left atrium; this foramen closes early but at the same time a second foramen forms called the interatrial foramen II; the atrial septum I contributes to the adult atrial septum.

atrial septum II
Figure 8.23
also see atrial septum I; interatrial foramen
A second atrial septum that forms later and on the right side of the atrial septum I; it overlaps the interatrial foramen II but it is also perforated by the foramen ovale; the atrial septa I and II fuse to form the adult atrial septum. The foramen ovale and interatrial foramen II do not close until after birth.

atrioventricular canal
Figures 8.22-8.24
The passage within the heart connecting the atrium with the ventricle; encircled and constricted by the endocardial cushion; it divides into right and left canals during the partitioning of the heart.

atrium
Figures 6.32, 6.44, 6.50, 6.60, 7.43-7.47, 7.56, 7.63, 7.64, 7.66-7.68, 7.70, 7.75, 7.84-7.88, 8.3, 8.5, 8.9, 8.20-8.24
also see left atrium; right atrium
The heart chamber that in the embryo receives blood from the sinus venosus and delivers it to the ventricle; in tetrapods, it becomes partitioned into right and left atria and incorporates the sinus venosus; in *Amphioxus,* it forms a chamber around the pharynx.

auditory nerve
see ganglion of cranial nerve VIII

auditory placodes
see otic placodes

auditory ganglion of cranial nerve VIII
Figures 8.6, 8.12
see ganglion of cranial nerve VIII

auditory vesicles
see otic vesicles

axial filament
Figures 1.11, 1.15
also see centrioles; sperm tails
A central fiber in cilia and the tail or flagellum of sperm; it is composed of a ring of nine double microtubules and a central pair of microtubules; it seems to arise from the centrioles near the head of the sperm; it functions in movements of the sperm tail.

basal plate
Figures 9.2-9.4
also see peripheral syntrophoblast; decidual cells; fibrinoid
A sheet of chorionic tissue in contact with the maternal decidua of the placenta; it is composed of peripheral syntrophoblast, peripheral syntotrophoblast and fibrinoid; it forms part of the afterbirth.

basement membrane
Figures 1.2, 1.5, 1.6, 7.2
A fibrous sheet beneath many kinds of epithelia; it supports the epithelia.

basilar artery
Figures 7.85, 8.6, 8.7, 8.11-8.14
A median vessel beneath the hindbrain; it connects the vertebral arteries with the internal carotid arteries.

bipinnaria larva
Figures 4.31, 4.32
A bilateral, free-swimming larval stage of starfish; after a growth period, the larva metamorphoses into an adult.

bivalent
Figure 2.2
A pair of homologous chromosomes in synapsis; found within the primary spermatocyte or primary oocyte during the 1st maturation division. *Synonym:* tetrad.

blastocoel
Figures 4.10-4.14, 4.25-4.29, 5.3, 5.12, 5.13, 5.19, 6.9-6.13, 6.16, 6.19
The cavity of the blastula; a closed space that is invaded early by mesenchyme and in some species is greatly reduced or obliterated by the enlarging archenteron. *Synonym:* segmentation cavity.

blastoderm
Figures 7.5-7.7
The nearly flat sheet of cells spreading over the surface of the yolk; it is present during an early stage in the development of telolecithal eggs (eggs of some fish, reptiles and birds), a stage that corresponds to the blastula stage of homolecithal eggs.

blastomeres
Figures 3.15, 4.6, 4.8, 4.22-4.24, 5.7, 5.8
also see founder cells
Cells formed during cleavage and composing the cleavage and blastula stages.

blastopore
Figures 3.12, 4.13, 4.28-4.30, 5.14, 5.15, 6.11, 6.12, 6.14-6.17
The opening of the archenteron to the exterior; it marks the point of origin of the archenteron and the caudal end of the embryo; it forms the anus in echinoderms.

blastula
Figures 4.10, 4.11, 4.26, 4.27, 6.10
The final stage of cleavage–typically, a hollow sphere–with the embryonic cells surrounding a cavity, the blastocoel.

blood islands
Figures 7.14, 7.20, 7.30, 7.42
A mass of splanchnic mesodermal cells in the gut wall of amphibians or the yolk sac of amniotes; the first blood forming tissue of the embryo, it forms red blood cells and the vitelline blood vessels.

blood vessels
Figures 1.1, 1.2, 1.7, 1.8, 7.5, 8.10
also see medulla; myometrium
Blood vessels extend into organs and spread through them following strands or sheets of connective tissue. They provide oxygen and nutrients to growing tissues and remove the waste products of metabolism.

body stalk
Figure 7.48, 8.29
The slender link between bird embryos and their extraembryonic membranes; it is covered by amnion and contains the vitelline and allantoic blood vessels; it is homologous to the umbilical cord of mammals.

body wall
Figures 7.43, 7.44, 7.47, 7.68, 7.76, 7.88, 8.20, 8.21, 8.25
The outer layer of the body, enclosing the body cavities and viscera; it derives from the epidermis, dermatome, myotome and somatic mesoderm of the embryo.

brachial plexus
Figures 8.28, 8.30
An interconnection of ventral rami of cervical and thoracic spinal nerves from which branch the nerves of the forelimb.

brain vesicle
Figures 5.4, 5.16, 5.17, 6.17, 6.57
also see neural tube
In *Amphioxus,* the slightly enlarged cranial end of the neural tube.

branchial arch 1
Figures 6.34, 6.43, 6.44, 6.47, 6.48, 7.43, 7.44, 7.53, 7.63, 7.64, 7.68-7.70, 7.73, 7.74, 7.87, 8.1, 8.3, 8.4, 8.16-8.19
also see aortic arch 1; mandibular process; maxillary process
The most cranial branchial arch, composed of a mandibular process forming the caudal border of the stomodeum, and a maxillary process cranial to the stomodeum; in early stages it contains aortic arch 1; its caudal boundary is branchial groove 1; it forms the lower jaw and sides of the upper jaw.

branchial arch 2
Figures 6.27, 6.31, 6.32, 6.43, 6.44, 7.53, 7.54, 7.63, 7.74, 7.84, 7.88, 8.4, 8.16, 8.18, 8.19
also see aortic arch 2
The second branchial arch; it arises as a thickening of the pharyngeal wall between branchial grooves 1 and 2; it contains aortic arch 2 in early stages; it forms the stapes (columella), styloid process, stylohyoid ligament, part of the hyoid bone, root of the tongue and the facial muscles.

branchial arch 3
Figures 7.88, 8.16
also see aortic arch 3
The third of a series of paired bars in the wall of the pharynx, within which are formed the third aortic arch and a cartilage bar of the visceral skeleton; in aquatic vertebrates, it forms and supports the gills; in mammals, it contributes to the epiglottis, and

in tetrapods, to the hyoid bone.

branchial arch 4
Figures 6.43, 7.73, 8.16
also see aortic arch 4
The fourth of a series of paired bars in the wall of the pharynx, within which are formed aortic arch 4 and a cartilage bar of the visceral skeleton; in aquatic vertebrates, it forms and supports the gills; in tetrapods, it contributes to the larynx, and in mammals, to the epiglottis.

branchial cleft 1
see branchial groove 1; pharyngeal pouch 1
A slitlike perforation in the wall of the pharynx between branchial arches 1 and 2; it is formed at the level of pharyngeal pouch 1 and branchial groove 1; it forms the Eustachian tube, middle ear cavity and external ear canal. *Synonym:* gill slit 1; hyomandibular cleft; visceral cleft 1.

branchial cleft 2
Figure 6.28
also see branchial groove 2; pharyngeal pouch 2
A slitlike perforation in the wall of the pharynx between branchial arches 2 and 3; it is formed at the level of pharyngeal pouch 2 and branchial groove 2; it subsequently closes and is obliterated. *Synonym:* gill slit 2; visceral cleft 2.

branchial cleft 3
Figure 6.28
also see pharyngeal pouch 3
A slitlike perforation in the wall of the pharynx between branchial arches 3 and 4; it is formed at the level of pharyngeal pouch 3 and branchial groove 3; it subsequently closes. *Synonym:* gill slit 3; visceral cleft 3

branchial cleft 4
see branchial groove 4; pharyngeal pouch 4
A thin region in the wall of the pharynx between branchial arches 4 and 5; it is formed at the level of pharyngeal pouch 4 and branchial groove 4; it forms only in lower vertebrates (frog). *Synonym:* gill slit 4; visceral cleft 4.

branchial groove 1
Figures 7.52, 7.53, 7.63, 7.67, 7.69, 7.70, 7.73, 7.74, 7.84, 8.4, 8.13-8.15
also see branchial cleft 1
An ectodermal invagination meeting pharyngeal pouch 1; it forms the external ear canal in amniotes. *Synonym:* visceral groove 1.

branchial groove 2
Figures 7.43, 7.44, 7.63, 7.64, 7.67, 7.69, 7.70, 7.74, 8.1, 8.18
also see branchial cleft 2
An ectodermal invagination meeting pharyngeal pouch 2; it is obliterated in tetrapods. *Synonym:* visceral groove 2.

branchial groove 3
Figures 7.44, 7.63, 7.67, 7.69, 7.70, 7.84, 7.87, 8.1
also see branchial cleft 3
An ectodermal invagination meeting pharyngeal pouch 3; it is obliterated in tetrapods. *Synonym:* visceral groove 3.

branchial groove 4
Figures 7.67, 7.70, 7.84, 8.17
also see branchial cleft 4
An ectodermal invagination meeting pharyngeal pouch 4; it is obliterated in tetrapods. *Synonym:* visceral groove 4.

branchial pouch
see pharyngeal pouch

C blastomere
see founder cells

cardiac jelly
Figure 7.29
A gelatinous substance between the endocardium and myocardium of the early embryonic heart; it maintains the separation between the heart layers for a period of time; it is eventually replaced by proliferating cells of the heart wall.

caudal arteries
Figures 6.57, 8.6, 8.7
The extension of the dorsal aortae into the tail.

caudal cardinal veins
Figures 6.43, 6.45, 6.52, 6.57, 7.33, 7.45, 7.47, 7.59, 7.60, 7.66, 7.67, 7.69, 7.75-7.80, 7.85-7.87, 8.2, 8.3, 8.7, 8.9, 8.25, 8.26, 8.28, 8.30-8.32, 8.34, 8.35
The primitive paired veins of the trunk; they lie dorsal to the mesonephros and drain with the cranial cardinal veins into the common cardinal veins; they form the iliac veins. *Synonym:* postcardinal veins.

caudal fin
Figures 5.1, 5.18
also see dorsal fin
The tail fin.

caudal intestinal portal
Figures 7.45, 7.62, 7.63, 7.81
also see midgut
The opening of the hindgut of amniotes; it moves cranially, lengthening the hindgut; it meets the cranial intestinal portal at the level of the small intestine, forming the yolk stalk.

caudal liver diverticulum
Figures 7.58, 7.77
also see cranial liver diverticulum; liver
One of two outgrowths of the duodenum of birds which grow, branch and anastomose to form the liver.

caudal veins
Figures 6.45, 6.57
The principal veins of the tail; they drain into the caudal cardinal veins.

caudal vitelline vein
Figure 7.46
also see vitelline veins A branch of the left vitelline vein extending caudally to receive the sinus terminalis.

cecum
Figures 8.5, 8.35
A pocket in the large intestine of mammals near its connection to the small intestine; it appears as an enlargement of the intestinal loop early in the development of the intestine.

celiac artery
Figures 7.85, 8.6, 8.7
A ventral branch of the aorta supplying the cranial digestive system; it derives from the vitelline arteries.

cell center
see centrosome

cell column
Figure 9.4
Tissue anchoring villi within the placenta.

centrioles
Figures 1.11, 1.15
also see sperm tails; axial filament
A pair of minute cytoplasmic granules, usually near the nucleus, surrounded by a zone of gelated cytoplasm; self-replicating and composed of a ring of nine sets of ultramicroscopic tubules; they are associated with the formation of spindle fibers and the axial filaments of cilia and flagellae.

centrosome
Figure 2.4
A cytoplasmic organelle composed of two minute granules, the centrioles and surrounding fibrous protein; it forms the mitotic spindle in dividing cells and in sperm, also the axial filament.
Synonym: cell center.

cerebral aqueduct
Figures 7.47, 7.68, 7.86, 8.8
also see fourth ventricle; third ventricle
The neural canal of the mesencephalon; the originally large chamber is progressively narrowed by the thickening of the walls of the midbrain, becoming a slender canal connecting the third and fourth ventricles.

cerebral hemispheres
Figures 7.70, 7.76, 7.84, 7.88, 8.5, 8.20, 8.21
also see telencephalon
Paired dorsolateral bulges of the telencephalon; they form the cerebrum.

cervical flexure
Figure 7.67
also see cranial flexure
One of several ventral bends in the body axis giving the amniote embryo a compact C-configuration; it forms in the region of the hindbrain and cranial trunk.

cervical sinus
Figures 8.4, 8.17
A depressed region in the sides of the neck bearing branchial grooves 3 and 4 in its floor; it subsequently closes and is obliterated.

chalaza
Figure 7.6
also see albumen
Strands of dense albumen present in the bird egg and attached to the vitelline membrane of the yolk; it holds the yolk near the center of the egg, yet allows the yolk to rotate and to float upward to a position near the surface of the upper shell, a position favorable for development during incubation.

chitinous layer
Figures 2.2-2.6
A thick, clear layer of the egg shell of *Ascaris;* formed from the egg after separation of the fertilization membrane; it is composed of chitin and protein.

chordamesoderm
Figures 6.12, 6.13, 6.15, 6.16
The archenteron roof of amphibian gastrulae; it arises by involution of marginal zone cells over the lips of the blastopore; it forms the notochord in the dorsal position and the adjacent mesoderm.

chorda-neural crescent
Figures 5.3, 5.19
also see mesodermal crescent
In *Amphioxus* embryos during cleavage, this is a zone on the dorsal side of the embryo opposite the mesodermal crescent; the part lying above the equator will subsequently form the nervous system, the part of the crescent below the equator will form the notochord after gastrulation.

chorioallantois
Figures 7.48, 7.89
A composite structure formed by the fusion of the chorion and allantois; the principal respiratory organ of bird embryos; it forms the albumen sac of bird embryos and absorbs the albumen; it forms the embryonic component of the placenta of most mammals; it is vascularized by the allantoic blood vessels.

chorion
Figures 7.48, 7.49, 7.51, 7.52, 7.55, 7.56, 7.60, 7.75, 7.78, 7.80-7.83, 7.89, 9.4
also see amniotic folds
The outermost embryonic membrane of amniotes; it arises from somatopleure and is usually drawn over the embryo by the amniotic folds; it later fuses with the allantois to form the chorioallantois, which is vascularized by the umbilical vessels and functions as the main embryonic respiratory organs of birds; the chorioallantois contributes to the placenta of mammals.

chorionic plate
Figure 9.4
also see placental villi
The basal layer of the chorionic component of the placenta from which the chorionic villi arise.

chromosomes
Figures 2.3, 2.7, 2.8
Threads of chromatin in the cell nucleus or in the mitotic and meiotic division figures; they contain the genes in linear order.

cicatrix
Figure 7.5
A thin vascular strip in the follicular wall of bird ovaries; rupture of the mature follicle to release the egg during ovulation occurs through the cicatrix.

ciliated band
Figures 4.16, 4.18, 4.19, 4.31, 4.32
Continuous strips of tall ciliated cells in the epidermis of echinoderm larvae (pluteus, bipinnaria) extending over the body and arms; it functions in locomotion and food gathering.

cleavage furrow
Figures 3.5, 5.9, 6.8, 6.9, 7.7
A constriction in the cytoplasm that divides the egg or blastomere; it forms during the telophase of the cleavage divisions; the constriction is formed by a contractile ring of actin microfilaments.

cleavage stage
Figures 2.5, 2.6, 4.6-4.11, 4.22-4.27, 5.7-5.13, 6.8-6.10, 7.7, 7.8
The period of development beginning with the first mitotic division of the egg and ending with the blastula; a period of rapid mitoses during which no growth occurs, consequently, the cells

becoming smaller as they become more numerous.

cloaca
Figures 3.1, 6.53, 6.55, 6.57, 6.65, 7.5, 7.63, 7.69, 7.83, 7.84, 7.88, 8.32, 8.35
also see urogenital sinus
The caudal chamber of the vertebrate digestive tract; it receives the allantoic stalk, urinary ducts and reproductive ducts; it is partitioned in mammals to form the rectum, urinary bladder and urogenital sinus.

cloacal membrane
Figures 7.83, 8.9, 8.32
A double-layered membrane formed where the ventral wall of the cloaca fuses with ventral ectoderm; it ruptures to open the anus and, in mammals, the urogenital sinus as well.

clubshaped gland
Figures 5.4, 5.17, 5.18
also see clubshaped gland duct
In *Amphioxus,* a temporary larval organ arising as an evagination of the gut between the endostyle and pharynx; it forms an external opening ventral to the mouth; it degenerates at the larval stage of 12 gill slits.

clubshaped gland duct
Figure 5.4
also see clubshaped gland
The duct is located just ventral to the mouth.

coelom
Figures 4.1, 6.38, 7.24, 7.25, 7.33, 7.36, 7.37, 7.40, 7.41, 7.47, 7.62, 7.68-7.70, 8.37
also see coelomic vesicles; embryonic coelom; extraembryonic coelom; intraembryonic coelom; pericardial cavity
A cavity within the mesoderm that forms the body cavities of the adult; in vertebrates, it arises as a cleft in the lateral plate mesoderm, which is thereby divided into somatic and splanchnic layers.

coelomic sac
Figures 4.15, 4.18, 4.19, 4.30-4.32
also see archenteric vesicle
A thin-walled, mesodermal evagination of the archenteron of echinoderm gastrulae (sea urchin, starfish); it forms the coelom and water vascular cavities of the adult.

collecting tubules
Figure 6.1
In the frog testis, conducting vessels carrying mature sperm from the seminiferous tubules to the vasa efferentia.

colon
Figures 8.6, 8.36, 8.37
also see cecum
Part of the large intestine; it arises from the hindgut.

common bile duct
Figures 8.31, 8.33
A vessel connecting the cystic and hepatic ducts to the duodenum; it conducts bile into the duodenum, and arises from the stem of the liver diverticulum.

common cardinal veins
Figures 6.45, 7.33, 7.45, 7.57, 7.66, 7.67, 7.69, 7.75, 7.85, 7.88, 8.2, 8.9, 8.21, 8.22, 8.24, 8.29

The trunk of the cranial and caudal cardinal veins; they connect to the sinus venosus and contribute to the superior vena cava and the oblique vein of the adult.

common carotid artery
Figure 8.7
The common vessel that branches into the internal and external carotid arteries.

common iliac arteries
Figures 7.85, 8.38
also see umbilical arteries
The large terminal branches of the aorta; they arise from the proximal segments of the umbilical arteries and a pair of dorsal intersegmental arteries; they persist as the common trunks of the external and internal iliac arteries.

common vitelline vein
Figure 8.35
An enlarged vein within the umbilical cord associated with the yolk sac.

connective tissue
Figures 1.7, 7.1, 7.4
One of the main kinds of tissue; characterized by much intercellular material, including fibers. *Synonym:* stroma.

conotruncus
Figures 6.32, 6.44, 6.45, 6.49, 6.57, 7.30, 7.32, 7.33, 7.36, 7.43, 7.44, 7.46, 7.47, 7.55, 7.56, 7.63, 7.64, 7.66, 7.68, 7.69, 7.75, 7.84, 7.87, 8.3, 8.6, 8.8
The heart chamber, originally most cranial in position, connecting the ventricle with the ventral aorta (and later, aortic sac); in tetrapods it is partitioned longitudinally to form the ascending aorta and pulmonary aorta.

copula
Figure 8.16
also see tongue
A median elevation on the floor of the mouth arising from branchial arch 2 and contributing to the root of the tongue.

coronary sinus
Figure 8.29
A venous trunk vessel within the dorsal wall of the of the heart which drains the coronary veins of the heart into the right atrium; it derives from a persistent left common cardinal vein.

corpus luteum (*pl.* corpora lutea)
Figure 1.9
A mass of endocrine gland tissue in the ovary of viviparous vertebrates; it forms from an ovulated follicle and persists into pregnancy; it secretes progesterone.

cortex
Figures 1.7, 1.9, 5.5
The outer zone of an organ or egg; it contains the follicles and corpora lutea of the ovary.

cranial cardinal veins
Figures 6.42, 6.45, 7.33, 7.45, 7.46, 7.50-7.56, 7.66, 7.67, 7.69, 7.71-7.74, 7.85, 7.87, 7.88, 8.2, 8.7, 8.9, 8.12-8.15, 8.17-8.20
The primitive, paired veins of the head and neck; they drain with the caudal cardinal veins into the common cardinal veins; they form the cerebral veins, dural sinuses, internal jugular veins and superior vena cava. *Syn:* precardinal veins.

cranial cartilages
Figure 6.59
Cartilage bars forming part of the primitive skull or chondocranium; they arise from mesodermal head mesenchyme and neural crest; they are eventually replaced by bony skull in higher vertebrates.

cranial flexure
Figure 7.67
also see cervical flexure; tail flexure
One of several ventral bends in the body axis giving the amniote embryo a compact C-configuration; it forms in the midbrain and is the only permanent flexure.

cranial intestinal portal
Figures 7.20, 7.24, 7.28-7.32, 7.38, 7.43-7.45, 7.59, 7.63, 7.64, 7.78
also see midgut
In amniotes, the opening from midgut into foregut; it moves caudally, lengthening the foregut, and meets the caudal intestinal portal at the level of the small intestine, forming the yolk stalk.

cranial liver diverticulum
Figures 7.44, 7.58, 7.76
also see liver
One of two outgrowths of the duodenum of birds that grow, branch and anastomose to form the liver.

cranial nerve III
Figures 7.64, 7.73, 7.74, 7.87, 8.6, 8.11, 8.12
A pair of motor nerves arising from the floor of the mesencephalon; they innervate some extrinsic and all intrinsic eye muscles. *Synonym:* oculomotor nerve.

cranial nerve IV
Figures 8.6, 8.10
A pair of motor nerves arising from the mesencephalon and emerging from the roof of the brain at the isthmus; they innervate the superior oblique ocular muscles. *Synonym:* trochlear nerve.

cranial nerve V
Figures 6.30, 6.35, 6.42, 6.45, 6.48, 6.60, 7.43, 7.44, 7.50, 7.51, 7.63, 7.64, 7.67, 7.69-7.73, 7.84, 7.87, 8.1, 8.2, 8.5, 8.6, 8.11, 8.13-8.15
also see ganglion of cranial nerve V; mandibular ramus of cranial nerve V; maxillary ramus of cranial nerve V; ophthalmic ramus of cranial nerve V; root of cranial nerve V; semilunar ganglion
A pair of mixed nerves arising from the sides of the metencephalon and semilunar ganglia; three divisions–the ophthalmic, maxillary and mandibular rami–innervate the mandibular arch region. *Synonym:* trigeminal nerve.

cranial nerve VI
Figure 8.13
A pair of motor nerves emerging from the floor of the myelencephalon; they innervate the external rectus eye muscles. *Synonym:* abducens nerve.

cranial nerve VII
Figures 6.30, 6.36, 6.42, 6.45, 6.49, 6.60, 7.43, 7.44, 7.52, 7.63, 7.64, 7.67, 7.72, 7.84, 7.87, 8.2, 8.6, 8.11-8.14
also see ganglion of cranial nerve VII; geniculate ganglion
A pair of mixed nerves arising from the myelencephalon at the cranial margin of the otic vesicle and the geniculate ganglion; they innervate branchial arch 2. *Synonym:* facial nerve.

cranial nerve VIII
Figures 6.30, 6.36, 6.45, 6.49, 7.43, 7.44, 7.52, 7.63, 7.64, 7.67, 7.69, 7.71, 7.84, 7.87, 8.2, 8.12
also see auditory ganglion; ganglion of cranial nerve VIII
A pair of sensory nerves arising from the auditory ganglion; they innervate the otic vesicle. *Synonym:* auditory nerve.

cranial nerve IX
Figures 6.30, 6.42, 6.45, 6.50, 6.61, 7.43, 7.44, 7.53, 7.54, 7.63, 7.64, 7.69, 7.71, 8.2, 8.6, 8.12-8.14
also see ganglion of cranial nerve IX; superior ganglion; petrosal ganglion. A pair of mixed nerves arising from the myelencephalon at the caudal margin of the otic vesicles; they bear the superior and petrosal ganglia and innervate branchial arch 3. *Synonym:* glossopharyngeal nerve.

cranial nerve X
Figures 6.30, 6.45, 6.50, 6.62, 7.54, 7.55, 7.64, 7.67, 7.69, 7.72, 8.2, 8.6, 8.11-8.15, 8.17
also see ganglion of cranial nerve X; jugular ganglion; accessory ganglia of cranial nerve X; nodose ganglion
A pair of mixed nerves arising from the myelencephalon and bearing the jugular and nodose ganglia; they innervate branchial arches 4, 5 and 6, and extend parasympathetic fibers to the viscera; in aquatic vertebrates (frog tadpole), they innervate the lateral line. *Synonym:* vagus nerve.

cranial nerve XI
Figures 7.67, 8.6, 8.11-8.13
A pair of motor nerves arising from the myelencephalon and spinal cord and innervating muscles of the pharynx and shoulder; part of the vagus nerve in aquatic vertebrates. *Synonym:* accessory nerve.

cranial nerve XII
Figures 8.6, 8.8, 8.12-8.15, 8.17
A pair of nerves arising from many rootlets on the ventral wall of the myelencephalon; they innervate the tongue muscles and they evolved from cervical spinal nerves of aquatic vertebrates. *Synonym:* hypoglossal nerve.

cranial neuropore
Figures 7.19, 7.20, 7.23, 7.32
A temporary cranial opening into the neural tube; it is obliterated by complete closure of the prosencephalon.

cranial venous plexus
Figures 7.33, 7.85, 8.7
The principal embryonic vein draining the venous plexi of the brain and later the dural sinuses; it is derived from the cranial segment of the cranial cardinal vein.

cranial vitelline veins
Figure 7.46
also see vitelline veins
Paired veins draining the cranial part of the yolk sac and the sinus terminalis; they empty into the right and left vitelline veins.

cumulus oophorus
Figure 1.8
also see stratum granulosa
A thickening in the stratum granulosa containing the mammalian oocyte.

cuticle
Figures 3.24, 3.25
In *C. elegans* embryos, the lining of the gut.

cystic duct
Figure 8.33
also see gallbladder
The duct of the gallbladder connecting it with the common bile duct; it arises from the liver diverticulum.

cytoplasm
Figures 4.1, 4.4, 4.5, 4.20, 5.5, 5.6
The part of the cell or egg (oocyte) excluding the nucleus; in eggs, it contains yolk.

cytotrophoblast
Figure 9.4
also see villus cytotrophoblast

D blastomere
see founder cells

decidual cells
Figures 9.2-9.4
Enlarged connective tissue cells of the maternal decidua; the decidua lies adjacent to the basal plate; it contains abundant stores of glycogen and lipid.

decidua of uterus
Figure 9.4
also see decidual cells; endometrium; placental villi
The inner superficial zone of the endometrium; during pregnancy it is in direct contact with the placental villi; it is detached during the contractions following birth and it is expelled as part of the afterbirth.

dense mesenchyme
Figures 8.14, 8.15
A condensation of mesenchymal cells preparatory to the formation of the eye muscles in these locations.

dermatome
Figures 6.51, 7.59, 7.75
also see somites
The outermost division of the somite; it lies under and in contact with epidermis forming the dermis or connective tissue of the skin.

descending aorta
Figures 7.46, 7.47, 7.56-7.59, 7.63, 7.64, 7.66, 7.68, 7.72-7.79, 7.86, 7.87, 8.3, 8.6-8.9, 8.28, 8.30-8.32, 8.34-8.38
also see dorsal aorta
The principal artery of the trunk; a median vessel formed by the fusion of the paired dorsal aortae; it extends from the subclavian to the common iliac arteries. *Synonym:* aorta.

diencephalon
Figures 6.46, 6.47, 6.56, 6.59, 7.43, 7.44, 7.47, 7.50-7.54, 7.63, 7.67, 7.69, 7.70, 7.75, 7.76, 7.84, 7.86, 7.87, 8.1, 8.3, 8.8, 8.13-8.15, 8.17, 8.18
The caudal division of the prosencephalon; a deep, laterally compressed region to which the optic stalks, infundibulum and epiphysis attach; its cavity is the third ventricle of the brain; it forms the epithalamus, thalamus and hypothalamus; its roof forms the choroid plexus.

differentiating spermatid
Figures 1.10, 1.15
also see spermatid
The spermatid during its transformation into a sperm, a process called spermiogenesis; during differentiation the spermatid is embedded in a pocket within the Sertoli cell. *Synonym:* immature sperm.

diffuse stage
Figure 7.3
A phase in the diplotene stage of meiosis in some species; the chromosomes become extended and diffuse resembling the interphase state.

diplotene stage
Figures 1.3, 1.14, 7.3
also see pachytene stage
A stage of the first maturation division in spermatogenesis and oogenesis that follows the pachytene stage; during the diplotene stage, chromosomes in synapsis separate except at points of crossing over (chiasmata); it is followed by diakinesis during which the chromosomes separate farther moving the chiasmata to the ends of the chromosomes (terminalization); metaphase of the first maturation division follows diakinesis.

dorsal
Figures 3.12, 3.17-3.19
The side of the embryo opposite the gut; the neural tube forms on the dorsal side of vertebrate embryos and *C. elegans* embryos.

dorsal aortae
Figures 6.45, 6.51-6.54, 6.56, 6.57, 6.65, 7.30-7.33, 7.35, 7.39, 7.45-7.47, 7.52-7.55, 7.60, 7.61, 7.66-7.68, 7.70, 7.72, 7.80, 7.83, 7.86-7.88, 8.14, 8.15, 8.17-8.22, 8.24-8.26
also see descending aorta
The primitive paired, longitudinal arteries of the trunk that fuse together caudal to the pharynx to form the descending aorta; in the pharyngeal region, they contribute to the external carotid arteries, descending aorta and, in mammals, the right subclavian artery.

dorsal diverticulum
Figures 5.4, 5.15
also see left diverticulum; right diverticulum
In *Amphioxus,* a dorsal evagination of the cranial end of the gut; it extends ventrally and divides into the left and right diverticula which separate from the gut.

dorsal fin
Figures 5.1, 6.39, 6.53, 6.54
A flat extension of the body wall along the dorsal midline of the trunk and tail; it degenerates during metamorphosis in anurans.

dorsal lip
Figures 5.3, 5.14, 5.19, 6.11-6.16
also see ventral lip; lateral lips
The margin of the blastopore toward the animal pole and at the dorsal side of the embryo; the first blastopore lip to form in amphibians, it derives from the gray crescent area of amphibian eggs; it forms the foregut roof, notochord and dorsal mesoderm; in *Amphioxus* it forms the notochord.

dorsal mesentery
Figures 7.58, 7.59, 7.76-7.78, 8.34, 8.35, 8.38
also see greater omentum
A double layer of splanchnic mesoderm suspending the gut from the dorsal body wall and extending from the esophagus to the cloaca; it provides a path and support for nerves and vessels of the gut; it forms the mediastinum, greater omentum and mesenteries of the intestine, and it contributes to the diaphragm.

dorsal mesocardium
Figures 7.29, 7.37, 7.55, 7.57, 7.75
The dorsal mesentery of the heart; it derives from the ventral

mesentery of the foregut; it soon ruptures and disappears.

dorsal pancreatic rudiment
Figures 7.77, 8.6, 8.32-8.34
also see ventral pancreatic rudiment
A dorsal evagination of the duodenum; together with a ventral evagination (two in frogs and birds), it forms the rudiments of the pancreas, which fuse to form one glandular mass.

dorsal root ganglia
see spinal ganglia

dorsal root of spinal nerve
Figures 8.28, 8.30, 8.39
also see spinal ganglia
The dorsal division of the spinal nerve connecting the trunk of the nerve to the alar plate of the spinal cord; it is composed of sensory nerve fibers; it bears the dorsal root ganglion. *Synonym:* sensory root.

ductus venosus
Figures 7.67, 7.68, 7.76, 7.86, 7.87, 8.7, 8.8, 8.28, 8.29
A vein in the liver of amniotes carrying blood from the vitelline and left umbilical veins to the sinus venosus; it derives from vitelline veins; it is obliterated after hatching in birds or after birth in mammals.

duodenum
Figures 7.58, 7.77, 8.32-8.34
The first segment of the small intestine; it arises from the foregut; it forms the diverticula of the liver and pancreas; it later connects with the common bile duct and pancreatic ducts.

dyad
Figure 2.7
An individual chromosome formed by the division of a tetrad.

E blastomeres (cells)
see founder cells

ectoderm
Figures 5.14-5.16, 5.20, 6.13, 6.15, 7.13, 7.22, 7.23, 7.25, 7.32, 7.35, 7.37, 7.42, 7.48
also see epidermal ectoderm; inner layer of ectoderm
The outermost of the three primary germ layers; it develops into the epidermis, skin glands, hair, feathers, nails, scales, nervous system, lining of the nose, inner ear, retina, lens of the eye, pituitary gland, mouth, pigmented cells and anus.

ectodermal fold
Figure 5.20
The *Amphioxus* equivalent of a neural fold.

eggs
see oocytes

egg pronucleus
Figures 4.1, 4.3
also see mature ova
The haploid nucleus of the mature egg formed by the completion of the second maturation division.

egg shell
Figure 2.5
also see shell
The outer protective covering of eggs; in *Ascaris* the shell consists

of an outer uterine layer derived from uterine fluid, a fertilization membrane and an inner chitinous layer derived from the egg.

embryonic shield
Figure 7.8
also see hypoblast
A shieldlike darkening of the caudal blastoderm during the pre-streak stage of chick embryos; it marks the caudal end of the future embryonic axis and consists of hypoblast.

EMS blastomere
see founder cells

endocardial cushion
Figures 8.2, 8.22-8.24
A pair of connective tissue outgrowths that undergo fusion to divide the atrioventricular canal; it contributes to the atrioventricular valves.

endocardial tube
Figure 7.28
also see endocardium; heart
The thin tubular lining of the embryonic heart; it eventually fuses with the myocardium to form the heart wall.

endocardium
Figures 7.28, 7.29, 7.37, 7.56, 7.57, 7.75, 7.76
The lining of the heart; arises from splanchnic mesoderm; it fuses with the myocardium later to form the heart wall.

endoderm
Figures 5.14, 5.15, 5.19, 5.20, 6.13, 6.15-6.20, 6.23, 6.24, 6.26-6.29, 6.31, 6.32, 6.40, 6.41, 6.43, 6.44, 7.12, 7.16-7.18, 7.22-7.24, 7.26, 7.27, 7.29, 7.35, 7.41, 7.42, 7.48, 7.79
also see hypoblast; yolk endoderm
The innermost of the three primary germ layers, inward movement of which is part of gastrulation; it forms the lining of the digestive and respiratory tracts, the pancreas, liver, thyroid, parathyroid glands, thymus, primordial germ cells (except in urodeles), bladder and urethra; and in *Amphioxus,* it forms the head cavity, preoral cavity, endostyle and clubshaped gland.

endolymphatic duct
Figure 8.12
The stalk of the otic vesicle; except in elasmobranchs, it soon loses its connection with the body surface; it forms part of the inner ear.

endometrium
Figure 9.1
also see uterine epithelium; uterine glands
The glandular membrane lining the uterus; it consists of uterine epithelium, glands and connective tissue stroma with blood vessels; it shows a marked sequence of changes during the menstrual cycle (growth, secretion, degeneration or menstruation and repair) which is regulated by ovarian hormones; it receives and nourishes the embryo during pregnancy.

endostyle
Figures 5.4, 5.17, 5.18
In *Amphioxus,* a ciliated groove in the floor of the pharynx; it secretes and transports mucus with food toward the esophagus; it arises as a thickening in the right cranial wall of the gut.

enterocoel
Figure 5.20
also see somite
The cavity of the somites as in *Amphioxus;* it is derived from the archenteron; it forms the myocoel, and ventrally the enterocoels of somites fuse to form coelom.

eparterial bronchus
Figures 8.24, 8.27
An unpaired evagination of the trachea which, together with the lung buds, form the lung; it subsequently forms the upper lobe of the right lung.

epiblast
Figures 7.9, 7.10, 7.12, 7.18
The upper or outer layer of the blastoderm of birds and mammals; in the area pellucida of birds the original caudal half of the epiblast migrates through the primitive streak to form endoderm and mesoderm; the cranial half of the epiblast becomes ectoderm.

epibranchial placodes
Figure 6.37
Ectodermal thickenings dorsal to the branchial grooves; they contribute cells to the cranial ganglia.

epidermal ectoderm
Figures 5.19, 6.16, 6.18, 6.46, 7.13, 7.20, 7.23-7.27, 7.29, 7.31, 7.33-7.36, 7.38-7.41, 7.45, 7.67
also see epibranchial placodes; lens placodes; olfactory placodes; otic placodes
The epithelial outer covering of the embryo; it mostly forms epidermis, but some placodes (thickenings) arise from it, which contribute to the sense organs and cranial ganglia.

epidermal growth
Figure 5.19
In *Amphioxus,* the spread of epidermal cells over the neural groove and blastopore; it results in formation of the neurenteric canal and neuropore.

epidermis
Figures 6.17, 6.19, 6.20, 6.23, 6.24, 6.26, 6.33
The outer, epithelial layer of the skin; it derives from the ectoderm.

epiglottis
Figures 8.15, 8.16
An elevation on the floor of the pharynx cranial to the glottis in mammals; it is composed of cartilage in the adult; it derives from branchial arches 3 and 4.

epiphysis
Figures 6.28-6.30, 6.40, 6.45, 6.46, 7.45, 7.63, 7.64, 7.67, 7.68, 7.84
An evagination from the roof of the diencephalon; it forms the pineal gland.

esophageal plug
Figure 6.50
A mass of cells temporarily blocking the esophagus before the amphibian larva begins to feed.

esophagus
Figures 4.15-4.18, 4.31, 4.32, 6.56, 6.62, 6.63, 7.47, 7.75, 7.86, 8.8, 8.22, 8.24-8.26
Part of the digestive tract that connects the pharynx (or mouth in the sea urchin) with the stomach; it lengthens markedly in amniotes during development; it arises from the foregut in vertebrates and from the archenteron in sea urchins; it forms the crop of birds.

external carotid arteries
Figures 6.57, 7.85, 8.7, 8.17
Arteries of the head; they arise as outgrowths of the aortic sac near the base of aortic arch 3; initially, they supply branchial arches 1 and 2.

external gill
Figure 6.43
also see gill
Respiratory organs growing out of branchial arches 3 to 6 in amphibia; they aerate blood from the aortic arches; later they are covered by the operculum; they are replaced by internal gills.

external jugular veins
Figures 6.45, 8.18
Veins of the head; they arise as branches of the cranial cardinal veins; initially, they drain branchial arches 1 and 2.

external layer
Figure 2.5
The outermost layer of the egg shell of *Ascaris;* it is adherent to the fertilization membrane and is formed from a uterine secretion.

external nares
Figure 6.55
also see internal nares; olfactory pit
The external opening of the nasal passage; derived from the olfactory pit.

extraembryonic coelom
Figures 7.34, 7.35, 7.38, 7.39, 7.42, 7.48, 7.50, 7.60, 7.61, 7.71, 7.75, 7.80, 7.82, 9.4
also see coelom
The division of the coelom outside the head fold of the body, the lateral body folds and the tail fold of the body; it lies between the chorion and amnion and between the chorion and yolk sac; in chick and pig, it receives the expanding allantois.

facial nerve
see cranial nerve VII

falciform ligament
Figure 8.31
The ventral ligament of the liver; it attaches the liver to the ventral body wall; it derives from the ventral mesentery.

fan
Figure 3.1
In *C. elegans* adult males, an appendage projecting from the tail.

fertilization membrane
Figures 2.2-2.4, 4.7, 4.21, 4.25, 5.2, 6.12
A membrane separated from the surface of the egg after fertilization in many aquatic species; it derives from the vitelline membrane and is often thickened by material from the cortical granules; it contributes to the egg shell in *Ascaris.*

fetal arteries
Figure 9.2
also see stem placental villi
In the placenta, branches of the umbilical arteries found in the

placental plate and in the villi; they supply the capillaries of the villi with fetal blood.

fetal blood cells
Figures 9.3, 9.4
also see fetal capillaries; fetal arteries
In the human placenta, the blood cells within the umbilical and placental vessels; they are separated from the maternal blood cells by the placental membrane except during labor when tears in the membrane may permit some mixing.

fetal capillaries
Figure 9.3
also see placental membrane
In the placenta, the anastomosing capillary bed of the placental villi; they contain fetal blood supplied by the umbilical arteries and drained by the umbilical vein; they form part of the placental membrane.

fibrinoid
Figure 9.2
also see basal plate
An intercellular matrix of the basal plate in which peripheral cytotrophoblast cells are frequently embedded; it gives a positive periodic acid-Schiff reaction for polysaccharide.

floating villi
Figure 9.4
also see placental villi
Free placental villi attached only at their base to the chorionic plate; they are suspended in the maternal blood of the intervillous spaces.

follicle cells
Figures 1.7, 6.4
also see stratum granulosa
The epithelial cells enclosing the oocyte; they probably regulate the transfer of materials to the oocyte; they form the stratum granulosa of bird and mammalian follicles.

follicular cavity
Figures 1.7-1.9
The space within the Graafian follicle; it is filled with a viscous follicular fluid. *Synonym:* antrum.

forebrain
see prosencephalon

foregut
Figures 6.17-6.19, 6.37, 6.43, 6.44, 6.51, 7.20, 7.21, 7.23, 7.24, 7.29-7.32, 7.43, 7.44, 7.57-7.59
also see pharynx
The part of the gut extending into the head from the midgut; in amniotes, it is formed by the head fold of the body as it passes under the head and trunk; it eventually forms the pharynx, respiratory tract excepting the nasal passages, esophagus, stomach, duodenum, liver and pancreas.

foreleg bud
Figures 8.1, 8.2, 8.4, 8.5, 8.22, 8.25, 8.26, 8.28, 8.30
also see hindleg bud; leg bud
The rudiment of the foreleg; it arises as a thickening of somatic mesoderm in the body wall; an ectodermal thickening, the apical ectodermal ridge, forms on its margin; it is homologous to the wing bud of birds and arm bud of humans.

founder cells
Figures 3.2-3.14
Certain blastomeres formed during cleavage of *C. elegans* embryos. They give rise to the major tissues of the adult, and they consist of 6 progenitor cells (AB, MS, E, C, D and P4). The ABa, ABp and right and left blastomeres, EMS, MSa, MSp, Ea, and Ep blastomeres, P0, P1, P2 and P3 blastomeres and Z2 and Z3 blastomeres are part of the lineage of founder cells.

fourth ventricle
Figures 7.47, 7.68, 7.86, 8.8
also see cerebral aqueduct
The enlarged neural canal of the rhombencephalon; it is connected cranially to the cerebral aqueduct and caudally to the central canal of the spinal cord; its thin roof forms the choroid plexus of the medulla.

Froriep's ganglion
Figures 8.6, 8.13
In some mammals, the most caudal of the accessory cranial ganglia; it may contribute to cranial nerve XII.

gallbladder
Figures 6.62, 8.6, 8.8, 8.32, 8.33
A saclike vessel connected by the cystic duct to the common bile duct; it arises from a caudal extension of the liver diverticulum; it stores bile.

ganglion of cranial nerve V
Figures 6.42, 6.48, 6.60, 7.44, 7.50, 7.51, 7.64, 7.67, 7.69-7.73, 7.84, 7.87, 8.1, 8.2
also see cranial nerve V; semilunar ganglion
Ganglion arising from the cranial neural crest and from the epibranchial placode above the first branchial groove; it is also from the dorsolateral placodes in lower vertebrates; it supplies sensory fibers to cranial nerve V. *Synonym:* semilunar ganglion; trigeminal ganglion.

ganglion of cranial nerve VII
Figures 6.42, 6.49, 6.60, 7.43, 7.44, 7.52, 7.63, 7.64, 7.67, 7.84, 7.87, 8.1, 8.2, 8.5, 8.13
also see cranial nerve VII; geniculate ganglion
Ganglion arising from the preotic neural crest and from the epibranchial placode above the first branchial groove; also from the dorsolateral placodes in lower vertebrates; it forms beside the ganglion of cranial nerve VIII and supplies sensory fibers to cranial nerve VII. *Synonym:* geniculate ganglion.

ganglion of cranial nerve VIII
Figures 6.49, 7.43, 7.44, 7.52, 7.63, 7.64, 7.67, 7.69, 7.71, 7.87, 8.2
also see auditory ganglion
Ganglion formed by aggregating cells detached from the otic placode and from the otic vesicle; it lies between the geniculate ganglion and the otic vesicle; it later divides into the spiral and vestibular ganglia of the inner ear; it supplies the fibers of cranial nerve VIII. *Synonym:* auditory ganglion.

ganglion of cranial nerve IX
Figures 6.42, 6.50, 6.61, 7.43, 7.44, 7.54, 7.63, 7.64, 7.67, 7.69, 7.71, 7.84, 8.2
also see cranial IX; superior ganglion; petrosal ganglion
Ganglion formed from the postotic neural crest and cells from the epibranchial placode above branchial groove 2; it is also from the dorsolateral placodes in lower vertebrates; it divides subsequently into the superior and petrosal ganglia and supplies sensory fibers

to cranial nerve IX.

ganglion of cranial nerve X
Figures 6.50, 6.62, 7.54, 7.55, 7.64, 7.67, 7.69, 7.72, 8.2
also see accessory ganglia; cranial nerve X; jugular ganglion; nodose ganglion
Ganglion formed from the postotic neural crest and cells from the epibranchial placode above branchial groove 3; it is also from the dorsolateral placodes in lower vertebrates; it divides subsequently into the jugular and nodose ganglia and supplies sensory fibers to cranial nerve X.

gastrocoel
Figures 5.3, 5.4, 5.19, 6.2, 6.11, 6.12
see archenteron

gastrula
Figures 4.12, 4.13, 4.28-4.30, 5.14, 6.11, 6.12, 7.12-7.18
The embryonic stage during which the primitive gut or archenteron is formed; the period following the blastula stage and during which extensive cell migrations form the primary germ layers.

geniculate ganglion
Figures 8.6, 8.12, 8.13
see ganglion of cranial nerve VII

genital ridge
Figures 7.79, 8.34, 8.35, 8.37, 8.38
A thickening of splanchnic mesoderm (germinal epithelium) and of the underlying mesenchyme on the medial edge of the mesonephros; in the early stages, it contains large primordial germ cells; it forms the testis or ovary (except that in female birds the right genital ridge fails to develop). *Synonym:* germinal ridge.

genital tubercle
Figures 8.2, 8.4, 8.5, 8.9, 8.32
An elevation on the ventral body surface of mammals cranial to the cloacal membrane; it enlarges into the phallus and eventually forms the penis of males and clitoris of females.

germinal epithelium
Figure 1.7
The epithelial opening of the adult ovary and the embryonic gonad; it is derived from the splanchnic mesoderm and primordial germ cells.

germinal vesicle
Figures 2.1, 2.7, 4.1, 4.4, 4.20, 6.4, 7.4
The much enlarged nucleus of the oocyte; during the prophase of the 1st maturation division, its membrane breaks, releasing most of its contents into the cytoplasm.

germ line cells
Figure 3.2
In *C. elegans* embryos, the cells derived from the P4 blastomere that give rise to eggs and sperm.

germ wall
Figures 7.9, 7.10
In bird embryos the outer circular zone surrounding the cellular blastoderm; a region of nuclear proliferation and cell organization which contributes cells to the margin of the epiblast and hypoblast.

gill
Figures 6.50, 6.51, 6.57, 6.60, 6.61

also see external gill
In amphibians, filamentous respiratory organs growing out of branchial arches 3 to 6; external gills arise first and then are covered by an operculum; these degenerate as they are replaced by internal gills which subsequently also degenerate during metamorphosis.

gill bars and slits
Figure 5.1
also see pharynx
In *Amphioxus,* the gill bars form a series of paired skeletal rods in the pharyngeal wall with gill slits in between; gill bars contain aortic arches and support the nephridia; they arise from the pharyngeal rudiment of the gut.

gill pouch
Figure 6.17
see pharyngeal pouch

gill slit
Figure 5.4
see branchial cleft

gland cells
Figure 3.2
In *C. elegans* embryos, some of the cells derived from the MS blastomere; they give rise to glandular tissue.

glandular oviduct
Figure 7.5
also see albumen
A large segment of the oviduct of birds located between the isthmus and the uterus; it secretes the albumen of the egg when under the stimulus of estrogen. *Synonym:* magnum.

glossopharyngeal nerve
see cranial nerve IX

glottis
Figures 6.61, 8.16, 8.18, 8.19
The opening from the pharynx into the trachea of early embryos or into the larynx of later embryos; in mammals, it acquires lateral borders (the arytenoid folds) from branchial arches 4 and 5.

Graafian follicle
Figures 1.7, 1.9
The ovarian follicle of mammals containing a follicular cavity; it derives from a primary follicle and either atrophies (undergoes atresia) or ovulates its oocyte; the Graafian follicle subsequently forms a corpus luteum.

greater omentum
Figures 8.28, 8.31
A saclike membrane attached to the greater curvature of the stomach of birds and mammals; it contains a cavity, the omental bursa; it derives from the dorsal mesentery of the stomach.

grinder
Figures 3.23-3.25
In *C. elegans* embryos, the portion of the gut between the pharynx and intestine.

gut
Figures 3.2, 3.24, 3.25, 5.4, 5.16, 5.20, 7.13, 7.22, 8.29
also see archenteron
The digestive tract; it derives from the archenteron.

head or head end
see anterior

head fold of the body
Figures 7.14, 7.21, 7.24, 7.36
A downward bend of membranes around the head that mark the boundaries of the embryonic area; it undercuts the head, separating it from the extraembryonic area to form the subcephalic pocket; it forms the ventral surface of the head and the foregut and it is continuous caudally with the lateral body folds.

head mesenchyme
Figures 6.18, 6.21-6.23, 6.35, 6.46-6.48, 7.13, 7.20, 7.24, 7.25, 7.29, 7.35-7.38, 7.49, 7.50, 7.53
also see dense mesenchyme
A loose tissue surrounding the brain and foregut, largely of mesodermal origin; it derives from the paraxial mesoderm cranial to the somites, the prechordal plate and the neural crest; it forms the following head structures: blood vessels, skull, head muscles and connective tissue.

head plexus
Figures 7.46, 7.66
A dense capillary network surrounding the brain cranial to the myelencephalon; it is supplied by the internal carotid arteries and drained by the cranial cardinal veins; some head vessels develop later from the plexus.

head process
Figures 7.14, 7.15
also see notochordal process
In amniotes, a band of mesodermal cells extending cranially from Hensen's node; the rudiment of the notochord.

heart
Figures 6.28, 6.29, 6.36, 6.40, 6.41, 6.55, 6.56, 8.1, 8.4, 8.29
also see atrium; conotruncus; sinus venosus; ventricle
In early stages, a tubular organ divided by constrictions into sinus venosus, atrium, ventricle and conotruncus; its wall is formed of an inner endocardial layer and an outer myocardium; it arises from paired heart tubes derived from the splanchnic mesoderm which fuse beneath the foregut; in air-breathing vertebrates, it becomes more or less completely divided longitudinally in the later stages to provide a separate pulmonary circulation.

heart mesoderm (rudiments or tube)
Figures 6.26, 7.20, 7.24, 7.29
also see heart
The heart-forming tissue; in amphibians, it arises from paired areas of the mesodermal mantle passing over the lateral lips of the blastopore, moving forward then turning ventrally to a position under the foregut; in birds it arises from splanchnic mesoderm of both amniocardiac vesicles adjacent to the cranial intestinal portal.

Hensen's node
Figures 7.12, 7.14, 7.16, 7.19, 7.20, 7.30, 7.31
also see primitive streak
The cranial thickened end of the primitive streak. *Synonym:* Primitive knot.

hepatic cecum
Figure 5.1
In *Amphioxus,* a blind outgrowth of the stomach extending cranially along the right side of the pharynx, which probably serves as a digestive gland; it arises in a late larval stage. *Synonym:* liver, midgut, cecum.

hepatic ducts
Figure 8.33
Small ducts conducting bile from the liver to the common bile duct; they arise from the hepatic diverticulum.

hepatic veins
Figure 8.29
A pair of veins that drain blood from the liver into the inferior vena cava.

hindbrain
see rhombencephalon

hindgut
Figures 6.17, 6.25-6.27, 6.29, 6.31, 6.32, 6.39, 6.40, 6.43, 6.44, 6.56, 6.64, 7.68, 7.81, 7.82, 8.2
also see cloaca
The caudal part of the embryonic gut; it extends from the midgut to the tail in amniotes; it is formed as the tail fold passes under the caudal trunk region; it forms successively tail gut, cloaca, colon and caudal small intestine; in amphibians, it forms the rectum.

hindleg bud
Figures 6.65, 8.1, 8.4, 8.5, 8.36, 8.38
also see foreleg bud; leg bud
The rudiment of the hindleg; it arises as a thickening of somatic mesoderm in the body wall; an ectodermal thickening, the apical ectodermal ridge, forms on its margin in amniotes; it is homologous to the leg bud of birds and humans.

Hofbauer cell
Figure 9.4
Large, vacuolated spherical cells within the stroma of chorionic villi; its function is uncertain but it may be phagocytosis.

hypoblast
Figures 7.9, 7.10
also see embryonic shield
The inner or lower layer of the blastoderm of birds and mammals; it lies between the epiblast and subgerminal cavity or between epiblast and yolk; it is formed by inward migration and aggregation of large yolky cells; it is replaced by endoderm derived from the primitive streak.

hypochordal rod
see subnotochordal rod

hypodermal cells
Figures 3.2, 3.3
In *C. elegans* embryos, the cells that form the outer surface (hypodermis or epidermis).

hypoglossal nerve
see cranial nerve XII

hypomere
see lateral plate mesoderm

hypophysis
Figure 6.57
also see adenohypophysis; infundibulum; Rathke's pouch
An endocrine gland beneath the hypothalamus; it derives from the infundibulum and Rathke's pouch or, in amphibians, from the infundibulum and a solid ingrowth from the stomodeum. *Synonym:* pituitary gland.

ileocolon ring
Figures 5.1, 5.18
In *Amphioxus,* a thickened heavily ciliated segment of the intestine.

iliac arteries
Figure 8.7
A pair of arteries supplying through its branches the lower abdomen and lower limbs.

immature sperm
Figure 1.15
also see differentiating spermatid
The sperm during its final stage of differentiation; some further elongation of the head with its dense nucleus and of the tail will occur.

inferior vena cava
Figures 6.57, 8.7-8.9, 8.25, 8.26, 8.28, 8.29, 8.31, 8.32, 8.34
The principal systemic vein of the trunk; it derives from several primitive paired veins, including the right vitelline, right subcardinal, right supracardinal and right caudal cardinal; it originally drains into the sinus venosus but is carried into the right atrium as the sinus venosus merges with the atrium.

infundibulum
Figures 6.26, 6.27, 6.29, 6.31, 6.35, 6.41, 6.45, 6.56, 7.30-7.34, 7.43-7.45, 7.47, 7.52, 7.64, 7.67, 7.68, 7.74, 7.84, 7.86, 8.3, 8.5, 8.8
also see hypophysis
A ventral evagination of the prosencephalon; it lies in the floor of the diencephalon and later in the hypothalamus; it subsequently forms the neural (caudal) lobe of the hypophysis.

inner layer of ectoderm
Figures 6.12, 6.15
also see outer layer of ectoderm
The inner layer of ectoderm in amphibians, covered by the outer layer; it thickens in the regions of the neural plate and ectodermal placodes.

interatrial foramen
Figures 8.2, 8.21
also see interatrial foramen I; interatrial foramen II; atrial septum
An opening in the atrial septum allowing blood to pass from the right to the left side of the heart in tetrapods; it closes soon after breathing begins, completing longitudinal division of the heart.

interatrial foramen I
Figure 8.23
also see atrial septum I
An opening in the atrial septum I allowing blood to pass from the right to left atria; as foramen I closes, near the endocardial cushion a new interatrial foramen II opens through the cranial region of the atrial septum I, allowing continued flow between the atria.

interatrial foramen II
Figure 8.23
also see atrial septum I
A second opening in the atrial septum I allowing blood to pass from the right to the left atria after closure of interatrial foramen I.

intermediate layer
Figures 8.18, 8.28, 8.30

The middle layer of the developing neural tube; it contains neuroblasts from the ventricular layer and their nerve fibers; it forms the gray matter.

intermediate mesoderm
see nephrotome

internal carotid arteries
Figures 6.57, 7.46, 7.52, 7.66, 7.67, 7.72-7.74, 7.85, 7.87, 8.6, 8.7, 8.13, 8.14
The main arterial supply to the brain; they arise as cranial extensions of the dorsal aortae; later they acquire additions from the dorsal aortae and aortic arch 3.

internal nares
Figure 6.55
also see external nares; olfactory pits
The inner opening of the nasal passages; they are formed when the olfactory pits extend downward and break through the roof of the mouth.

interphase
Figures 1.3, 7.3
The period in the cell cycle between mitoses; in proliferating tissues, a time of synthesis of DNA, RNA and protein.

intersegmental arteries
Figures 7.58, 7.66, 8.6, 8.17, 8.39
Originally small paired branches of the dorsal aortae arising between the somites; they contribute to the following arteries: vertebrals, subclavians, intercostals and lumbars.

intersomitic furrows
Figure 7.20
also see somites
The spaces separating somites; the first to form lies between the 1st and 2nd somites; the others form within the segmental plate mesoderm in a craniocaudal sequence; they are subsequently obliterated by fusion of the adjacent somites.

interstitial cells
Figures 1.1, 1.2, 1.5, 1.7, 6.2
In the testis, clusters of endocrine gland cells (Leydig cells) between the seminiferous tubules which secrete testosterone; in the ovary, clusters of gland cells scattered in the cortex which derive from atretic follicles.

interventricular foramen
Figures 8.2, 8.22, 8.23
also see ventricular septum
An opening in the ventricular septum allowing blood to cross between the right and left ventricles in tetrapods; it closes during division of the conotruncus and the atrioventricular canal.

interventricular septum
Figures 8.22-8.25
A muscular partition arising from the caudal wall of the primitive ventricle; it grows cranially, fuses with the endocardial cushion and bulbar septum and divides the ventricle into the right and left ventricles.

interventricular sulcus
Figure 8.22
A longitudinal groove on the surface of the primitive ventricle marking the plane of its impending division into right and left ventricles.

intervillous space
Figures 9.2-9.4
also see placental villi
In the placenta, the region filled with maternal blood and in which
the placental villi are suspended; maternal blood supply to the
intervillous space is by way of the spiral arteries of the basal
plate; veins of the basal plate return the blood to the maternal
circulation.

intestinal loop
Figures 8.2, 8.8, 8.9, 8.31, 8.32, 8.35
also see cecum; colon; small intestine
A ventral extension of the intestinal loop into the umbilical cord
of mammals bearing an enlargement, the cecum; coiling soon
retracts it into the peritoneal cavity.

intestine
Figures 3.1, 3.18, 4.15-4.19, 4.31-5.1, 5.16-5.18, 6.55-6.57, 6.61-
6.64, 8.3
also see cecum; colon; duodenum; intestinal loop; rectum
Segment of gut following the stomach; it derives from both the
foregut and hindgut in amniotes, from the midgut as well in
amphibians and from the archenteron in *Amphioxus,* sea urchin
and starfish; the intestinal lining develops from gut endoderm;
muscle, connective tissue, blood vessels and serosa develop from
splanchnic mesoderm.

intraembryonic coelom
Figures 7.38, 7.39, 7.43, 7.44, 7.47, 7.57, 7.58, 7.60, 7.61, 7.79,
7.82, 7.86, 7.88
also see coelom; extraembryonic coelom
The division of the coelom within the body folds; it forms the
pericardial, pleural and peritoneal cavities of the adult.

isthmus
Figures 7.5, 7.43, 7.44, 7.47, 7.49, 7.63, 7.64, 7.67, 7.68, 7.70,
7.71, 7.86, 8.3, 8.8, 8.10
also see glandular oviduct
A narrow part of the oviduct of birds where the shell membranes
of the egg are formed; the narrowed connection between the
cerebral aqueduct of the mesencephalon and the 4th ventricle of
the metencephalon; thickening of the floor of the brain
progressively reduces the isthmus.

jugular ganglion
Figures 8.2, 8.6, 8.11
also see ganglion of cranial nerve X
A large ganglion of the cranial nerve X dorsal to the nodose
ganglion; it contributes sensory fibers to cranial nerve X.

junctional complex
Figure 1.6
An intercellular organelle of epithelial cells which binds the cells
together and forms a barrier against passage of substances
between cells; it consists of the zonula adherens, zonula occludens
and gap junctions.

lampbrush chromosomes
Figure 6.4
Enlarged chromosomes of oocytes with many lateral loops which
extend the length of the chromosome during the period of intense
RNA transcription and yolk synthesis.

laryngotracheal groove
Figures 7.47, 7.56, 7.68, 7.74
A trough in the floor of the caudal pharynx from which arise the
lung buds; it also contributes to the larynx and trachea.

larynx
Figure 8.20
The voice box; it derives from the upper trachea, the floor of the
pharynx and branchial arches 4 and 5.

lateral body folds
Figures 7.37, 7.48, 7.60, 7.61, 7.78, 7.79
A depressed fold in the somatopleure and splanchnopleure on
each side of the embryonic trunk; together with the head and tail
fold of the body, to which it connects, it forms the boundary
between the intraembryonic and extraembryonic regions.

lateral lingual swellings
Figures 8.15, 8.16
also see tongue
Paired elevations on the mandibular processes within the mouth;
they fuse with the tuberculum impar to form the body of the
tongue.

lateral lips
Figures 6.14, 6.16
also see dorsal lip; ventral lip
The lips on the right and left rims of the blastopore formed by
extensions of the dorsal lip; in amphibians, the midgut endoderm,
lateral plate and heart mesoderm pass over the lateral lips during
gastrulation.

lateral nasal processes
Figure 8.20
Elevations on the embryonic face lateral to the olfactory pits; they
fuse with the maxillary and medial nasal processes and form the
sides of the nose.

lateral plate mesoderm
Figures 6.16, 6.19, 6.20, 6.24, 6.27, 6.37, 6.45, 6.52, 7.13, 7.26,
7.27, 7.32, 7.39, 7.41, 7.45
also see ventral mesoderm; somatic mesoderm; splanchnic
mesoderm
The mesodermal layer lateral and ventral to the nephrotome; it is
split by the coelom into the somatic and splanchnic mesoderm.

lateral transverse vein
Figure 8.7
In the pig embryo, part of the plexus of small veins draining the
mesonephros.

lateral ventricles
see third ventricle; telencephalon
Lateral extensions of the original third ventricle of the
telencephalon into the cerebral hemispheres; they retain their
connections to the third ventricle of the diencephalon through the
foramen of Monroe; the thin roof projects into the lumen as a
choroid plexus.

left atrium
Figures 8.20-8.22, 8.24
also see atrial septum; atrium
The left division of the primitive heart atrium in tetrapods; it is
separated from the right atrium by the atrial septum; it receives
blood through the interatrial foramen and the pulmonary veins
before breathing begins; after closure of interatrial foramen, it is
supplied by increased pulmonary flow.

left diverticulum
Figures 5.4, 5.16, 5.17
also see dorsal diverticulum; preoral pit
A ventral thick-walled extension of the dorsal diverticulum; an

external opening, the preoral pit, is near the mouth.

left horn of sinus venosus
Figures 8.22, 8.24
also see sinus venosus
The part receiving the left common cardinal, left vitelline and left umbilical veins; it conducts the flow of blood to the transverse sinus venosus whence it passes into the right atrium through the sinoatrial opening; as the venosus return is directed to the right horn of the sinus venosus, the left horn is reduced, forming finally the oblique vein of the left atrium.

left umbilical vein
see umbilical vein

left ventricle
Figures 8.21, 8.22, 8.24, 8.25
also see ventricle
Heart chamber formed from the partitioning of the primitive ventricle by the ventricular septum; it receives blood from the left atrium and delivers it under high pressure to the ascending aorta.

leg bud
Figures 7.63, 7.67-7.69, 7.81, 7.82, 7.84, 7.86-7.88
also see foreleg bud; hindleg bud
The rudiment of the leg; it arises as a thickening of somatic mesoderm of the body wall, later bearing an ectodermal thickening, the apical ectodermal ridge.

lenses
Figures 6.55, 6.59
also see lens placode; lens vesicle
Induced by the optic vesicles from the overlying ectoderm; they are eventually enclosed by a lens capsule.

lens placodes
Figures 6.28, 6.34, 7.34
also see lens vesicles
A thickening of the head ectoderm overlying each optic vesicle; they invaginate to form the lens vesicles and subsequently the eye lenses.

lens vesicles
Figures 6.42, 6.47, 7.43-7.45, 7.53, 7.63, 7.64, 7.67, 7.75, 8.5, 8.17
also see lens placodes
An ectodermal sac within each optic cup; they derive from the lens placodes and forms the eye lenses.

leptotene stage
Figures 1.3, 1.12, 7.3
also see prochromosome stage
An early stage of the first maturation division in spermatogenesis and oogenesis; chromosomes have the form of separate long thin threads except that the X-chromosome may be a dense contracted body; it is followed by the zygotene stage.

lesser omentum
Figures 8.28, 8.31
A membrane that attaches the lesser curvature of the stomach to the liver; it derives from the ventral mesentery of the stomach.

lip
Figures 6.55-6.58
Extensions of skin on the upper and lower jaws of amphibian tadpoles forming the border of the mouth.

liver
Figures 6.55, 6.57, 6.62, 6.63, 7.67, 7.87, 8.1-8.4, 8.8, 8.9, 8.26
also see liver diverticulum; liver sinusoid; cranial liver diverticulum; caudal liver diverticulum
The largest of the digestive glands; it is important in fetal life for blood homeostasis and blood formation; it arises as a ventral diverticulum of the foregut in amphibians and mammals, and as two buds on the duodenum in birds; the buds branch and anastomose around the ductus venosus.

liver diverticulum
Figures 6.15, 6.19, 6.24, 6.26, 6.28, 6.29, 6.32, 6.37, 6.40, 6.41, 6.45, 6.51, 7.68
also see cranial liver diverticulum; hepatic ducts; liver; caudal liver diverticulum
The rudiment of the liver, gallbladder, hepatic ducts and common bile duct; it arises as a ventral evagination of the foregut in amphibians and mammals, and as two buds on the duodenum of birds.

liver sinusoids
Figure 8.28
The smallest blood vessels of the liver; they differ from capillaries in that their walls contain phagocytes; they derive originally from the ductus venosus and vitelline veins.

lumbosacral plexus
Figure 8.39
An anastomosis of spinal nerves in the caudal trunk region that supplies the nerves of the hind limb.

lungs
Figures 6.57, 6.62, 6.63, 8.27
also see eparterial bronchus; laryngotracheal groove; lung buds
They arise as a ventral diverticulum of the pharynx which branches repeatedly to form the endodermal lining of the trachea, bronchi and lungs; the mesoesophagus and lining of the pleural cavities form the mesodermal parts–muscle, connective tissue and pleura; the pulmonary arteries arise from aortic arch 6, and invade the lung; the pulmonary veins grow in from the atrium.

lung buds
Figures 6.45, 7.67, 7.70, 7.75, 7.87, 8.3, 8.6, 8.26
also see laryngotracheal groove
The paired rudiments of the lungs and bronchi; they arise from the laryngotracheal groove of the pharynx.

lymph sinus
Figures 6.4, 6.59
A large, lymph-filled space that drains into the veins; alternatively, the cavity of the hollow amphibian ovary.

macromeres
Figures 4.9, 5.10-5.12, 6.9, 6.10
The largest blastomeres or cells formed during cleavage; they form endoderm and in the sea urchin, ventral ectoderm also; in *Amphioxus,* they form endoderm, notochord and mesoderm.

mammary ridge
Figures 8.31, 8.32, 8.35
The rudiment of the mammary gland; it initially appears as an ectodermal thickening extending longitudinally between the bases of the limb buds; at the site of the mammary glands, the mammary ridges form tubular ingrowths, the milk ducts; renewed development of the mammary glands at puberty in females is a response to ovarian hormones; the development of the mammary gland is completed during pregnancy.

mandible
see mandibular arch; mandibular process
The lower jaw; it is formed by the fusion of the mandibular process of branchial arch 1.

mandibular process of branchial arch 1
Figures 7.43, 7.44, 7.53, 7.64, 7.68-7.70, 7.73, 7.74, 7.87, 8.1, 8.3, 8.4, 8.17-8.19
also see mandible; branchial arch 1
The caudal division of the mandibular arch; it forms the mandible, Meckel's cartilage, body of tongue, malleus, salivary glands and jaw muscles.

mandibular ramus of cranial nerve V
Figures 8.14, 8.15
also see cranial nerve V
The caudal division of cranial nerve V; it innervates the mandible and jaw muscles.

marginal layer
Figure 8.30
The outer layer of the developing neural tube; it contains neuroblasts derived from the inner layers and nerve fibers; it forms the white matter.

marginal zone
Figures 6.10, 6.11, 6.13, 6.14
The part of the animal hemisphere nearest the vegetal hemisphere; it turns in during gastrulation to form mesoderm and foregut endoderm.

maternal blood cells
Figure 9.3
also see intervillous space
In the living placenta, maternal blood cells fill the intervillous spaces, but they are mostly drained out in microscopic preparations of the placenta; maternal blood cells are separated from the fetal blood cells by the placental membrane.

maturation division I
Figures 1.2, 1.10, 1.13, 2.2, 6.2, 6.3, 7.2, 7.3
The first of two specialized cell divisions during the formation of sperm and eggs; its prophase is long and includes synapsis, chromosome replication and crossing over; this division reduces the chromosome number to the haploid state; it forms the secondary spermatocytes of males and secondary oocyte and a polar body in females. *Synonym:* meiosis I, reduction division I.

maturation division II
Figures 1.10, 1.13, 2.3, 6.3
The second of two specialized cell divisions during the formation of sperm and eggs; it begins immediately after the first maturation division but in many eggs is not completed until after fertilization; no chromosome replication occurs; it forms the spermatids of males and the second polar body and mature egg in females. *Synonym:* reduction division II, meiosis II.

mature ova
Figure 4.1
also see oocyte; primary oocyte; egg pronucleus
The female germ cells after completion of the maturation divisions; they are derived from oocytes; in sea urchins, they are found in the lumen of the ovary where they may be fertilized.

maxillary process of branchial arch 1
Figures 7.63, 7.64, 7.69, 7.70, 7.73, 7.74, 7.87, 8.4, 8.17-8.19
also see mandibular arch
The cranial division of branchial arch 1; it forms the cheek, lateral part of upper jaw, palate and incus.

maxillary ramus of cranial nerve V
Figure 8.15
also see cranial nerve V
The middle division of cranial nerve V; it innervates the upper jaw and face.

medial nasal processes
Figure 8.20
Elevations of the embryonic face medial to the olfactory pits; the two medial nasal processes fuse to form medial part of the upper jaw.

medial transverse vein
Figure 8.7
In the pig embryo, part of a plexus of small veins draining the mesonephros.

medulla
Figures 1.7-1.9
The inner or deep division of an organ; in the ovary, a region of connective tissue and blood vessels lacking follicles.

meiosis
see maturation division

mesencephalon
Figures 6.21, 6.26, 6.28, 6.29, 6.34, 6.40, 6.41, 6.45, 6.47, 7.30, 7.31, 7.33, 7.35, 7.43-7.45, 7.47, 7.49, 7.63, 7.64, 7.67-7.73, 7.84, 7.86, 7.88, 8.1, 8.3-8.5, 8.8-8.12
The middle primary vesicle of the brain; it forms the visual and auditory centers (corpora quadrigemina in mammals) and motor centers for movements of the head; its cavity narrows to form the cerebral aqueduct; it bears cranial nerves III and IV. *Synonym:* midbrain.

mesenchyme
Figure 4.29
also see head mesenchyme; primary mesenchyme; secondary mesenchyme
Loosely scattered cells which in early development spread through the blastocoel; it may derive from any germ layer but in sea urchins it arises from micromeres and archenteron, and in starfish from the archenteric vesicle; in these groups, it constitutes part of the early mesoderm and forms skeleton.

mesentery
Figures 8.36, 8.37
A supporting membrane attached to the organs within the coelom; it carries the vascular and nerve supply of organs; it is formed by the fusion of the two layers of splanchnic mesoderm.

mesoderm
Figures 5.19, 5.20, 6.15, 6.16, 6.23, 7.13, 7.15-7.18, 7.22
also see primary mesenchyme; secondary mesenchyme; coelomic vesicles; head mesenchyme; lateral plate mesoderm; dorsal mesoderm; segmental plate mesoderm; somite; nephrotome; ventral mesoderm
The middle primary germ layer; it is formed by gastrulation movements; it forms dermis, muscle, skeleton, blood vessels, blood (excepting possibly lymphocytes), connective tissues, kidneys, ureters, gonads (excepting germ cells), reproductive tracts and peritoneum.

mesodermal crescent
Figures 5.3, 5.19
also see chorda-neural crescent
A light zone visible in the egg and during cleavage of *Amphioxus* embryos, located on the ventral side of the vegetal hemisphere; during gastrulation it divides and migrates dorsally to positions on either side of the notochord; it then separates from the archenteron wall to form the somites, which constitute the entire mesoderm in *Amphioxus.*

mesodermal groove
Figure 5.3
also see enterocoel
A dorsal extension of the archenteron formed by the evagination of the somites; it will form the enterocoel.

mesomeres
Figure 4.9
also see nephrotome
The cells of intermediate size that compose the animal hemisphere of sea urchin embryos during cleavage; they form the ectoderm of the gastrula and larva; they form also the nephrotome of vertebrates.

mesonephric ducts
Figures 6.64, 7.60, 7.61, 7.78, 7.79, 7.81, 7.83, 7.88, 8.6, 8.8, 8.9, 8.34-8.38
The excretory ducts of the mesonephros; they formed initially as the pronephric ducts by the caudal growth of the pronephric buds to the cloaca; they contribute to the metanephros of amniotes by forming one of its rudiments on each side, the ureteric bud; they mostly degenerate in female amniotes, but in males they form the ductus epididymis, ductus deferens and seminal vesicles; they form the ducts of the adult kidney (opisthonephros) of amphibians. *Synonym:* Wolffian duct.

mesonephric glomeruli
Figures 8.2, 8.6, 8.8, 8.9, 8.32, 8.34, 8.38
Tufts of capillaries within the Bowman's capsules which together form the renal corpuscles of the mesonephros; similar structures form in the metanephros; glomeruli of the pronephros are associated with the coelom rather than with Bowman's capsule.

mesonephric ridge
Figure 7.79
also see mesonephros
A bulge or thickening extending into the dorsal part of the embryonic coelom at midtrunk levels; it is formed by the growing mesonephros.

mesonephric tubules
Figures 7.60, 7.61, 7.79, 7.87, 8.1, 8.6, 8.8, 8.9, 8.30, 8.32, 8.34, 8.38
The kidney tubules of the mesonephros; they posses glomeruli and a well-formed, coiled tubular structure; they are excretory during the embryonic period of amniotes; most degenerate but some form the efferent ductules of male amniotes; they form adult kidney tubules in amphibians.

mesonephros
Figures 6.64, 7.44, 7.63, 7.68, 7.80, 7.86, 8.3-8.5, 8.28, 8.31, 8.35
also see mesonephric tubules; mesonephric glomeruli; mesonephric ridge; mesonephric duct
The second or middle kidney of amniotes; it contains well formed tubules with glomeruli that produce urine during the embryonic period; the arrangement of the tubules is not segmental; the pronephric duct is appropriated as the mesonephric duct; mostly

degenerates in the adult amniote except that in males the caudal parts form the male reproductive tract (efferent ductules, ductus epididymis, ductus deferens and seminal vesicles). *Synonym:* Wolffian body.

mesovarium
Figure 1.9
The mesentery of the ovary; it provides support for the ovary.

metanephric diverticulum
Figure 8.6
also see pelvis of metanephros; ureter
A rudiment of the metanephros; it arises as an outgrowth of the mesonephric duct near its junction with the cloaca; the stalk becomes the ureter and the expanded distal end forms the pelvis and collecting ducts of the metanephros. *Synonym:* ureteric bud.

metanephric duct
see ureter

metanephrogenic mesenchyme
Figure 8.37
A strand of dense mesenchyme surrounding the pelvis of the metanephros; it is continuous cranially with the mesonephrogenic tissue; it derives from the caudal nephrotomes; it forms the metanephric tubules. *Synonym:* metanephrogenic tissue.

metanephrogenous tissue
see metanephrogenic mesenchyme

metanephros
see metanephric diverticulum; metanephrogenic mesenchyme; pelvis of metanephros; ureter
The last of three pairs of kidneys to form in the amniotes; metanephric tubules arise from the caudal end of the nephrogenic cord (metanephrogenic mesenchyme); the metanephric duct and pelvis are formed from the metanephric diverticulum (ureteric bud), an outgrowth of the mesonephric duct; it begins functioning in the embryo and it continues as the adult kidney.

metaphase
Figures 1.13, 2.5, 2.6, 5.2, 7.3
also see anaphase; prophase; telophase
A phase in cell division in which chromosomes are aligned on the metaphase plate.

metencephalon
Figures 6.56, 7.43-7.45, 7.47, 7.49, 7.63, 7.64, 7.67-7.71, 7.84, 7.86, 7.88, 8.1-8.5, 8.8-8.13
The cranial division of the rhombencephalon; its roof expands greatly to form the cerebellum whereas the pons forms in its roof; nerve centers (nuclei) for several cranial nerves develop within, including those of cranial nerves V, VI, VII and VIII; its cavity becomes the 4th ventricle of the brain.

micromeres
Figures 4.9, 4.10, 5.10, 5.11, 6.9, 6.10
The smallest cells in cleavage stages; they lie near the vegetal pole in the sea urchin and migrate into the blastocoel to form the primary mesenchyme; they compose the entire animal hemisphere of the amphibian embryo, forming ectoderm and mesoderm after gastrulation; they compose the animal hemisphere of *Amphioxus* where they form ectoderm.

midbrain
see mesencephalon

midgut
Figures 3.17, 3.22, 6.17, 6.20, 6.24, 6.26, 6.27, 6.29, 6.31, 6.38, 6.52, 7.25-7.29, 7.39, 7.60, 7.70, 7.79
also see cranial intestinal portal
In amphibians the middle part of the gut with a small lumen and thick yolky floor; it derives from the archenteron and will form the small intestine; in amniotes, it is the middle part of the gut whose floor is the cavity of the yolk sac (yolk-filled in reptiles and birds); it is steadily diminished by the lengthening of the foregut and hindgut to a mere yolk stalk attached to the small intestine.

mitochondria
Figures 1.11, 4.2
Filamentous or globular membranous organelles found in all eukaryotic cells; they have characteristic internal membranous folds or cristae; they contain enzymes of the Krebs cycle which transfer energy released by oxidative phosphorylation to ATP.

mitochondrial body
Figure 1.11
A fused mass of mitochondria found in the spermatids of some insects.

mitosis
Figure 1.12
Cell division (replication).

mitotic figure
Figure 6.9
also see cleavage division
The mitotic apparatus, consisting of chromosomes, spindle fibers and centrioles; it appears during each cleavage division; it is often indistinct in preparations used for class work.

motor root
see ventral root of spinal nerve

mouth
Figures 3.18. 4.15-4.19, 4.31, 4.32, 6.55, 7.86, 8.15
also see oral evagination; stomodeum
The cranial opening of the digestive tract; it is derived partly from an ectodermal invagination on the ventral side of the head, the stomodeum; the endodermal rudiment arises from the cranial wall of the foregut and is for a time separated from the stomodeum by the oral plate; rupture of the membrane opens the mouth.

MS blastomeres
see founder cells

muscle
Figure 6.59
also see myotome
Skeletal muscles of vertebrates mostly arise from myotomes but some head muscles (those of the jaws and eyes) arise from head mesenchyme; the limb muscles also arise from myotomes (in fishes, fin muscles also arise from myotomes).

muscle cells
Figure 3.2
In *C. elegans* embryos, some of the cells derived from the AB, MS, C and D blastomeres; they give rise to muscular tissue.

muscle fibrillae
Figure 5.20
also see myotomes
Contractile filaments in developing muscle cells of the myotome.

myelencephalon
Figures 6.56, 6.60-6.63, 7.43-7.45, 7.47, 7.49-7.56, 7.63, 7.64, 7.67, 7.68, 7.70, 7.71, 7.84, 7.86, 7.88, 8.1, 8.3-8.5, 8.8-8.14
The caudal division of the rhombencephalon and the most caudal part of the brain; a transition region from brain to spinal cord; contains nerve centers (nuclei) of cranial nerves IX-XII; cranial nerves IX and X are attached to its sides, and in amniotes, cranial nerves XI and XII as well; its cavity becomes part of the 4th ventricle of the brain, and its roof, the choroid plexus; it forms the medulla.

myocardium
Figures 7.28, 7.29, 7.32, 7.37, 7.56, 7.57, 7.75, 7.76
The outer layer of the heart, including the heart muscle; it forms from splanchnic mesoderm; it eventually fuses with the endocardium to form the heart wall.

myometrium
Figure 9.1
also see endometrium; smooth muscle
The muscle layer of the uterus; it is composed of bands of smooth muscle and numerous blood vessels; myometrial contractions transport sperm into uterine tubes, and at the end of pregnancy, myometrial contractions produce the labor contractions.

myotome
Figures 5.1, 6.51, 6.56, 6.64, 7.59, 7.75, 8.17, 8.20, 8.31, 8.38-8.40
also see somites
The division of the somite that forms skeletal muscle of the body wall and limbs; it lies at first in the dorsal region of the somite then migrates ventrally under the dermatome; its cells elongate longitudinally forming a muscle segment; myotomes are comparatively large in amphibian embryos.

nasal cavity
Figure 6.57
also see olfactory pits
A canal extending from the nostril to the mouth; it is formed from the olfactory pits as they deepen and break through the roof of the mouth; it forms the olfactory organ. *Synonym:* nasal passage.

nasal pits
see olfactory pits

nasal placodes
see olfactory placodes

nephrotome
Figures 6.24, 7.39
also see pronephros
A stalk-like connection between the somite and lateral plate mesoderm; it forms segmental buds in the cranial region which hollow out to form the pronephric tubules and ducts; at caudal levels, it forms the mesenchyme that develops into tubules of the mesonephros and metanephros; it contributes to the gonads. *Synonym:* mesomere; intermediate mesoderm.

neural canal
Figure 5.20
also see neural tube
The cavity of the neural tube; in many embryos it temporarily opens externally through a neuropore and connects with the archenteron through the neurenteric canal. *Synonym:* central canal; neurocoel.

neural crest
Figures 6.17-6.19, 6.23, 6.30, 6.34-6.36, 7.34, 7.38

also see ganglion of cranial nerve V, etc.; spinal ganglia
An ectodermal mesenchyme arising from the neural folds; it aggregates in many locations to form cranial ganglia, spinal ganglia, autonomic ganglia and the adrenal medulla; it attaches to the epidermis to form pigment cells, and to the neural tube to form the pia mater; it forms the neurolemma sheath cells of nerves; some neural crest migrate into the branchial arches to form the visceral skeleton.

neural ectoderm
Figures 5.19, 6.16
also see chorda-neural crescent; neural crest; neural plate
That part of the ectoderm that normally develops into neural structures or tissue.

neural folds
Figures 6.14, 6.15, 6.18-6.20, 7.19, 7.20, 7.22, 7.23, 7.25, 7.26, 7.30, 7.41
The elevated edges of the neural plate; opposite neural folds are brought together during bending and closing of the neural groove; they fuse and bud off streams of neural crest cells.

neural groove
Figures 5.3, 5.4, 6.18-6.20, 7.23, 7.25, 7.26, 7.29, 7.41
also see neural plate; neural folds; neural tube
A trough formed by the bending or rolling up of the neural plate; it closes with formation of the neural tube.

neural plate
Figures 5.3, 5.20, 6.14, 6.15, 6.19, 6.20, 7.13-7.17, 7.19, 7.20, 7.25, 7.27, 7.29
also see neural folds; neural groove; neural tube; prosencephalon; rhombencephalon; spinal cord
The earliest rudiment of the central nervous system; it rolls up into the neural tube during neurulation.

neural tube
Figures 5.1, 5.16-5.18, 5.20, 6.17, 7.20, 7.22, 7.24, 7.29, 7.32, 7.33, 7.45
also see neural folds; neural groove; neural groove; prosencephalon; rhombencephalon; spinal cord
The tubular rudiment of the central nervous system formed from the neural plate during neurulation; its cranial part forms the brain and its remainder forms the spinal cord; in amphibians, its extreme caudal end contributes to the tail somites.

neurenteric canal
Figures 5.3, 5.19, 6.25
A temporary connection between the caudal end of the neural groove or neural tube and the archenteron or yolk sac cavity; it occurs in many embryos, including frog and man.

neurocoel
see neural canal

neuromere
Figures 7.45, 7.71
A unit of the brain formed by transverse constrictions; neuromeres are particularly prominent in the myelencephalon; they later vanish.

neuronal cells
Figure 3.2
In *C. elegans* embryos, some of the cells derived from the AB, MS and C blastomeres; they give rise to neuronal tissue.

neuropore
Figure 5.4

also see neural canal
The temporary cranial or caudal external opening of the neural canal.

nodose ganglion
Figures 8.6, 8.15
also see ganglion of cranial nerve X
A large ganglion of cranial nerve X lying ventral to the jugular ganglion; it contributes sensory fibers to cranial nerve X.

notochord
Figures 5.1, 5.3, 5.4, 5.16-5.20, 6.15-6.17, 6.19, 6.20, 6.22-6.26, 6.28-6.30, 6.35-6.42, 6.45, 6.48-6.57, 6.60-6.65, 7.20, 7.21, 7.24-7.27, 7.29, 7.30, 7.32, 7.33, 7.35-7.40, 7.45, 7.47, 7.51-7.54, 7.64, 7.67, 7.68, 7.72, 7.73, 7.80, 7.86, 7.87, 8.6, 8.14, 8.39
also see notochordal process
The axial skeleton of chordate embryos; it arises in amphibians and *Amphioxus* from cells turning in over the dorsal lip of the blastopore, and in amniotes from cells extending cranially from Hensen's node; it shows marked elongation; it underlies the midline of the neural tube from the mesencephalon to the end of the spinal cord; its gelatinous cells acquire a tough sheath forming a flexible skeletal rod; in higher vertebrates, it is small and is later replaced by the vertebral column.

notochordal process
see notochord
In amniotes, a band of mesodermal cells extending cranially from Hensen's node; the rudiment of the notochord. *Synonym:* head process.

nuclear envelope
see nuclear membrane

nuclear membrane
Figures 4.4, 4.20, 6.4
Encloses the substance of the nucleus; at the ultrastructural level, it is double-layered with pores; it is destroyed during prophase of mitosis and re-formed during telophase. *Synonym:* nuclear envelope.

nucleolus
Figures 4.4, 4.20, 5.5, 6.4
One or more dense spherical granules in the nucleus of many cells; it is composed of ribosomal RNA (rRNA) and protein; it disappears during mitosis to re-form in connection with a pair of nucleolar chromosomes; it is large in cells with high rates of protein synthesis; multiple nucleoli occur in amphibian oocytes as a result of rRNA gene amplification.

nucleus
Figures 3.15, 4.2, 5.5, 5.7, 6.4, 6.8
also see germinal vesicle; nuclear membrane; nucleolus; pronucleus
A large body within the cell during interphase; it contains the chromosomes and often one or more nucleoli; it is enclosed by a nuclear membrane; it disappears during mitosis and is the site of synthesis of most of the DNA and RNA of the cell.

oculomotor nerve
see cranial nerve III

olfactory organ
Figure 6.58
also see nasal cavity; olfactory pits; olfactory placodes
A tubular passage derived from olfactory pits which extends

inward and opens through the roof of the mouth as internal nares; olfactory neurons arise from the part of the organ that extends olfactory nerve fibers into the telencephalon. *Synonym:* nasal passage.

olfactory pits
Figures 6.28, 6.31, 6.33, 6.40, 6.43, 6.45, 6.46, 7.63, 7.64, 7.67, 7.76, 7.84, 7.88, 8.4, 8.5, 8.9, 8.20, 8.21
also see nasal cavity; olfactory placodes; olfactory organs
Cavities on the lateral surfaces of the head cranial to the eyes; they arise by invagination of the olfactory placodes; they deepen and break through the roof of the mouth in air breathers to form the nasal cavities; olfactory cells differentiate from their walls as do the olfactory nerves. *Synonym:* nasal pit.

olfactory placodes
Figures 6.27, 7.55
also see olfactory pits
Paired ectodermal thickenings on the lateral surfaces of the head cranial to the eyes; they invaginate to form the olfactory pits; they form the rudiments of the nasal passages. *Synonym:* nasal placodes.

omental bursa
Figures 8.28, 8.31
The cavity of the greater omentum; it arises as an invagination into the dorsal mesentery of the stomach; it is connected with the peritoneal coelom; two bursae form in birds.

oocytes
Figures 1.7, 3.1, 6.4
also see primary oocytes
Immature eggs, distinguishable from the other ovarian cells by their gigantic size and prominent nucleus (germinal vesicle); each grows from the smaller oogonium and becomes a mature egg upon completion of its growth and two maturation divisions.

opercular chamber
Figures 6.60-6.62
also see opercular opening
Gill chambers formed by a membranous outgrowth of branchial arch 2 of tadpoles, extending caudally over the gills and branchial clefts; the two chambers connect ventrally and open through an external spiracle on the left side. *Synonym:* branchial chamber.

opercular opening
Figure 6.57
A persistent excurrent opening at the caudal edge of the operculum on the left side of the larval frog body; it is closed during metamorphosis by proliferation of tissue that fills the opercular cavity. *Synonym:* spiracle.

ophthalmic ramus of cranial nerve V
Figure 8.14
also see cranial nerve V
The cranial division of cranial nerve V; it distributes sensory fibers to the facial region.

optic cups
Figures 6.28, 6.34, 6.40, 6.42, 6.45, 6.47, 7.43-7.45, 7.53, 7.63, 7.67, 7.75, 7.84, 8.1, 8.5, 8.15, 8.17
also see sensory layer of optic cup
Each consists of a double-walled chamber formed by invagination of an optic vesicle; a lens vesicle lies in its "mouth;" the optic vesicle remains connected to the diencephalon by the optic stalk; its outer wall forms the pigmented epithelium of the retina; its inner wall forms the sensory layer of the retina and the optic nerve

fibers which grow through the stalk to the brain; the rim of the cup contributes to the iris and ciliary body; the optic cup of mammals is much smaller than that of birds. *Synonym:* eye cups.

optic fissure
Figures 7.45, 7.54, 7.67, 8.2
A groove in the ventral wall of the optic cup and optic stalk; after mesenchyme and blood vessels invade the fissure, its lips fuse.

optic stalks
Figures 6.34, 6.47, 7.44, 7.54, 7.70, 7.75, 8.88, 8.9, 8.17
The narrow connection of the optic cup on each side to the diencephalon; it guides the growing optic nerves from the optic cup to the brain.

optic vesicles
Figures 6.21, 6.27, 6.31, 7.30, 7.32-7.34
also see optic cups
Each is a lateral evagination of the prosencephalon; by invagination of its outer wall it forms the optic cup and subsequently the retina. *Synonym:* eye vesicles.

oral arms
Figures 4.15-4.18
also see anal arms
Slender, paired extensions of the dorsal body wall of pluteus larva; they are supported by skeletal rods and bear bands of cilia; they function to stabilize and propel the larva and to collect food.

oral evagination
Figures 6.29, 6.32, 6.34, 6.41, 6.44
The endodermal rudiment of the mouth; an cranial evagination of the pharynx toward the stomodeum in amphibians; contact with the stomodeum forms the oral membrane, which subsequently ruptures to open the mouth.

oral field
Figures 4.31, 4.32
also see oral lobe
In starfish embryos, a depressed ventral area surrounding the stomodeum or mouth.

oral hood
Figures 5.1, 5.18
In *Amphioxus,* a thin-walled funnel-like structure of the head leading to a mouth; it is fringed by stiff tentacles or cirri.

oral lobe
Figures 4.16, 4.30-4.32
also see oral field
In starfish embryos, a projecting area above the oral field and separated from it by the preoral ciliary band.

oral membrane
Figures 7.23, 7.35, 7.47, 7.73
also see stomodeum
A double-layered membrane composed of the floor of the stomodeum and the cranial wall of the pharynx; rupture of the membrane opens the mouth into the pharynx.

ostium of oviduct
Figure 7.5
The opening of the oviduct into the peritoneal cavity near the ovary; it accepts the egg into the oviduct after ovulation.

otic placodes
Figures 6.22, 7.31, 7.37
also see otic vesicles

Each is a thickening of the head ectoderm lateral to the myelencephalon; it invaginates forming the otic pit, and separates from the ectoderm as the otic vesicle; it subsequently forms the inner ear and contributes cells to the ganglion of the cranial nerve VIII. *Synonym:* auditory placodes; ear placodes.

otic vesicles
Figures 6.28, 6.30, 6.36, 6.40, 6.42, 6.49, 6.55, 6.57, 6.60, 6.61, 7.43-7.45, 7.52, 7.63, 7.64, 7.67, 7.69, 7.71, 7.84, 7.87, 8.1, 8.2, 8.5, 8.6, 8.11-8.13
also see otic placodes
Each is a closed chamber formed by the invagination of the otic placode; it separates from the head ectoderm and subsequently forms the inner ear. *Synonym:* auditory vesicles; ear vesicles.

outer layer of ectoderm
Figure 6.12
also see inner layer of ectoderm
The outer layer of ectoderm in amphibians, which covers the inner layer.

ovarian lumen
Figure 4.1
The cavity within the ovary of sea urchins into which eggs are ovulated; it leads to the exterior through the genital pore.

ovarian wall
Figure 4.1
Connective tissue capsule enclosing the germ cells and accessory cells in the ovary of sea urchins.

ovary
Figures 1.7-1.9, 3.1, 4.1, 6.4, 7.4
The female gonad; the organ where mature eggs form and ovulate; in vertebrates, it also secretes the female sex hormones estradiol and progesterone.

oviduct
Figures 3.1, 7.5
The organ of the female's reproductive system that transports the egg from the ovary to the uterus or cloaca. In birds, the layers of the egg, such as the albumen, shell membranes and shell, are added to the ovum within the oviduct.

P0 blastomere (zygote)
see founder cells

P1 blastomere
see founder cells

P2 blastomere
see founder cells

P3 blastomere
see founder cells

P4 blastomere
see founder cells

pachytene stage
Figures 1.3, 1.12-1.14, 6.57, 7.3
also see diplotene stage
A stage of the first maturation division in spermatogenesis and oogenesis it follows the zygotene stage; cells during the pachytene stage each contain a haploid number of bivalents (tetrads) or double chromosomes in synapsis; bivalents shorten and thicken during pachytene; the pachytene stage is followed by the diplotene stage.

pancreas
Figures 6.57, 7.44
also see dorsal pancreatic rudiment; ventral pancreatic rudiment
A digestive and endocrine gland arising as outgrowths of the duodenum and liver diverticulum.

pancreatic duct
Figure 8.34
also see dorsal pancreatic rudiment; pancreas; ventral pancreatic rudiment
The tubular connection between pancreatic tissue and the duodenum; it is formed by the outgrowth of both the dorsal and ventral pancreatic rudiments; usually only the ventral duct persists.

parathyroid gland rudiments
Figure 8.17
Masses of endocrine gland tissue derived from pharyngeal pouches 3 and 4; they migrate to the vicinity of the thyroid.

paraxial mesoderm
Figure 6.19
The mesoderm lateral to the neural tube and notochord that forms the somites.

pectinate muscles
Figure 8.23
Muscles forming ridges on the inner surface of the atrial walls of the heart; their presence marks the primitive part of the atria.

pericardial cavity
Figures 6.28, 6.29, 6.32, 6.36, 6.41, 6.44, 6.49, 6.50, 6.60, 7.28, 7.29, 7.31, 7.33, 7.36, 7.37, 7.55-7.59, 7.75, 8.19-8.21, 8.24-8.26, 8.28
also see intraembryonic coelom
The large coelomic space around the heart; formed by a cleft in the lateral plate mesoderm of the head; part of the splanchnic mesodermal layer thus formed develops into heart; it is cut off from the pleuroperitoneal coelom by the pleuropericardial membranes.

peripheral cytoplasm
Figure 7.4
In the chicken oocyte, an outer or cortical layer of finely granular cytoplasm; as the oocyte matures it contributes to the blastodisc.

peripheral syntrophoblast
Figures 9.2, 9.3
also see villus syntrophoblast; placental villi; basal plate
An extension of the villus syntrophoblast covering the fetal surface of the basal plate; like villus syntrophoblast, it is in contact with maternal blood and underlain by cytotrophoblast.

peritoneal cavity
Figures 6.51, 6.64, 7.76, 7.77, 8.30, 8.38
also see intraembryonic coelom
The body cavity of the abdomen; it derives from the caudal region of the coelom; after the pericardial cavity is cut off from the pleuroperitoneal cavity, the latter is split into two pleural cavities and a peritoneal cavity by the pleuroperitoneal membranes.

perivitelline space
Figures 2.3-2.5, 4.21-4.23, 5.2
The space between the fertilization membrane and the egg surface; it lies between the zona pellucida and the egg in mammals and contains perivitelline fluid, the "culture medium" of the developing egg.

petrosal ganglion
Figures 8.6, 8.14
also see ganglion of cranial nerve IX
The more ventral of two ganglia of cranial nerve IX; it contributes sensory fibers to the nerve. *Synonym:* inferior ganglion.

pharyngeal membrane
see oral plate

pharyngeal pouches
Figure 6.50
also see pharyngeal pouches 1-4
Paired evaginations of the lateral pharyngeal wall; each one meets the corresponding ectodermal invaginations (branchial grooves) and lies between the branchial arches.

pharyngeal pouch 1
Figures 6.23, 6.27, 6.31, 6.35, 6.43, 7.52, 7.72, 8.6, 8.14
also see branchial groove 1
Paired endodermal evaginations of the lateral pharyngeal wall caudal to branchial arch 1; they extend dorsally toward the otic vesicles to form the Eustachian tubes and tympanic cavities. *Synonym:* branchial pouch 1; gill pouch 1; visceral pouch 1.

pharyngeal pouch 2
Figures 6.27, 6.43, 7.54, 7.72, 7.73, 8.6, 8.15
also see branchial groove 2
Paired endodermal evaginations of the pharyngeal wall caudal to branchial arch 2; they are subsequently obliterated, except that they contribute to the thymus in lower vertebrates (frog). *Synonym:* branchial pouch 2; gill pouch 2; visceral pouch 2.

pharyngeal pouch 3
Figures 6.27, 6.31, 6.36, 6.43, 7.55, 7.72, 7.74, 8.2, 8.15, 8.17
also see branchial groove 3
Paired endodermal evaginations of the pharyngeal wall caudal to the third branchial arch; they contribute to the thymus and parathyroid glands. *Synonym:* branchial pouch 3; gill pouch 3; visceral pouch 3.

pharyngeal pouch 4
Figures 6.31, 7.73, 7.74, 8.18
also see branchial groove 4
Paired endodermal evaginations of the lateral pharyngeal wall caudal to the fourth branchial arch; they contribute to the parathyroid glands. *Synonym:* branchial pouch 4; gill pouch 4; visceral pouch 4.

pharynx
Figures 3.1, 3.14, 3.17, 3.22, 3.23, 5.1, 5.16-5.18, 6.22, 6.23, 6.26-6.29, 6.31, 6.32, 6.35, 6.36, 6.40-6.43, 6.47-6.49, 6.55, 6.56, 6.58-6.61, 7.35-7.37, 7.45, 7.47, 7.52-7.56, 7.67, 7.68, 7.72, 7.74, 7.88, 8.3, 8.8, 8.9, 8.17, 8.19, 8.20
also see foregut
The region of the embryonic foregut bearing pharyngeal pouches; it is large in *Amphioxus,* fishes and amphibian tadpoles, and it forms the gills; in air breathers, it extends caudally to the glottis, and is much reduced and its pouches become transformed into other structures.

pia mater
Figure 8.11
The inner layer of the meninges; a delicate membrane on the brain and spinal cord derived from head mesenchyme and neural crest.

pigmented cortex
Figures 6.8, 6.9
The surface coat of amphibian eggs and early embryos; a gelled layer containing much melanin in the animal hemisphere.

pigmented layer of optic cup
Figures 6.47, 6.59, 7.75, 8.17
also see optic cups
The outer wall of the optic cup; it is formed from the medial half of the optic vesicle; it forms the pigmented epithelium of the retina, ciliary body and iris.

pigment spot
Figure 5.4
Pigment cells associated with the neural tube in *Amphioxus;* a light-sensitive organ.

pituitary
see hypophysis

placental membrane
Figure 9.3
also see placental villi
The layers of the placenta interposed between maternal blood and fetal blood; it includes in humans at least the chorionic endothelium of fetal capillaries and the syntrophoblast, but in some areas, also the cytotrophoblast and chorionic mesenchyme; it controls the exchange between maternal and fetal blood.

placental villi
Figures 9.2, 9.3
also see anchoring villi; stem placental villi; peripheral syntrophoblast; villus cytotrophoblast; fetal capillaries
Branching treelike outgrowths of the chorion into the maternal blood of the intervillous spaces; they are covered by two epithelial layers, the outer syntrophoblast and inner cytotrophoblast; they contain a mesenchyme connective tissue and fetal blood vessels; some villi attach to the maternal decidua; they form the placental membrane which is interposed between the maternal and fetal bloods, controlling the exchange of substances between the two blood streams.

placode-derived ganglionic cells
Figure 7.51
also see ganglion of cranial nerve V
Those ganglionic neuroblasts derived from ectodermal thickenings–the placodes; other neuroblasts arise from the neural crest.

pleural cavities
Figures 7.75, 8.24, 8.25
The body cavities surrounding the lungs; they derive from the craniodorsal divisions of the pleuroperitoneal coelom, which become isolated from the pericardial and peritoneal cavities.

pluteus larva
Figures 4.15-4.19
A bilateral, free-swimming larval stage of sea urchins, sand dollars and brittle stars; it possesses long ciliated arms; after a growth period, the larva becomes immobile and metamorphoses into an adult.

polar bodies
Figures 4.21, 5.19
also see polar body I; polar body II

polar body I
Figures 2.3-2.7
A small cell separated from the primary oocyte by the first maturation division; it may divide again but then degenerates.

polar body II
Figure 2.4, 2.8, 3.3
A small cell separated from the secondary oocyte by the second maturation division; it degenerates.

polar spindle
Figures 2.7, 5.2
The mitotic apparatus, consisting of microtubules and associated proteins.

portal vein
Figures 8.2, 8.6, 8.29, 8.32, 8.34
The vessel that carries blood from the superior mesenteric and splenic veins to the ductus venosus and sinusoids of the liver; it derives from parts of the right and left vitelline veins.

postanal gut
Figure 8.6
The extension of the hindgut into the tail; it gradually degenerates. *Synonym:* tail gut.

posterior
Figures 3.12, 3.14, 3.17-3.21
The tail end of the embryo.

prechordal plate
Figures 6.15-6.18, 6.26, 7.23
A mass of midline cells cranial to the notochord and beneath the prosencephalon; originally forming the midline cranial portion of the foregut, it contributes to the cranial head mesenchyme.

preoral gut
Figures 7.67, 7.68, 7.73, 7.86, 8.2
The projecting tip of the foregut cranial to the oral plate; it gradually atrophies. *Synonym:* Seessel's pouch.

preoral pit
Figure 5.4
also see left diverticulum
The external opening into the left diverticulum in *Amphioxus;* it is located cranial to the mouth and arises as an invagination of ectoderm.

presumptive fate map
Figures 5.19, 6.16, 6.17, 7.13
A graphic representation of the location of organ-forming regions drawn on an earlier embryonic stage–often on the late blastula or early gastrula stages. *Synonym:* prospective fate map.

primary follicles
Figures 1.7-1.9, 7.5
Small follicles of the mammalian ovary, each with only one layer of follicle cells surrounding the oocyte; they grow in response to follicle-stimulating hormone produced by the hypophysis.

primary mesenchyme
Figures 4.11-4.13
A loose cluster of cells near the vegetal pole and within the blastocoel of sea urchins; it derives from micromeres and contributes to the skeleton of the pluteus.

primary oocytes
Figures 1.7, 1.8, 2.1, 4.1, 7.4, 7.5
also see oocytes
Immature eggs prior to completion of the 1st maturation division.

primary spermatocytes
Figures 1.2, 1.3, 1.5, 1.10, 1.12-1.14, 6.2, 7.2, 7.3
Large germ cells of the testis formed by growth of spermatogonia;

they undergo the first maturation division to form secondary spermatocytes.

primitive groove
Figures 7.14, 7.17-7.20, 7.30
also see primitive streak
A depressed trough between the primitive folds; a region of ingression of epiblast cells into the mesoderm and endoderm.

primitive ridges
Figures 7.14, 7.17, 7.18, 7.20
also see primitive streak
The thickened ridges of the primitive streak; they are formed by the convergent flow of epiblast. *Synonym:* primitive folds.

primitive streak
Figures 7.12, 7.19, 7.20, 7.30, 7.31
A longitudinal thickening in the epiblast of early amniote embryos; it is formed by the convergent flow of epiblast toward the caudal midline; it is the site of ingression of epiblast cells into the mesoderm and endoderm and consists of parallel longitudinal ridges (primitive folds), separated by a primitive groove, and a cranial thickening, Hensen's node.

prism larva
Figure 4.14
The larval stage of sea urchins, sand dollars and brittle stars immediately preceding the pluteus larva stage.

proamnion
Figures 7.14, 7.20, 7.23, 7.34
A crescent-shaped area lacking mesoderm around the head of early bird embryos; initially it delimits the cranial end of the embryo; later it is drawn under the head by the head fold of the body and is partially invaded by the mesoderm.

prochromosome stage
Figure 1.12
also see leptotene stage
The earliest prophase stage of the first maturation division of spermatogenesis of some insects; during this stage, chromosomes contract into discrete bodies of which there are a diploid number; unraveling of the prochromosomes leads to the next or leptotene stage.

proctodeum
Figures 6.26, 6.28, 6.29, 6.39, 6.54
An ectodermal invagination on the ventral side of the trunk at the base of the tail; it breaks into the hindgut to form the anus. *Synonym:* anal pit.

programmed cell death
Figures 3.14, 3.17, 3.19
The naturally occurring death of cells during embryogenesis.

pronephric ducts
Figures 6.45, 6.52, 6.53, 6.57
also see mesonephric ducts; pronephros
A pair of ducts connecting the pronephros with the cloaca; they arise by caudal growth of the pronephric buds; they subsequently become the mesonephric ducts.

pronephric tubules
Figure 6.42
also see nephrotome; pronephric ducts; pronephros
Tubules derived from the cranial nephrotomes, one pair per segment; each bears a nephrotome opening into the coelom on the

distal end and its proximal end connects to a pronephric duct; they degenerate before the adult stage but are functional during the larval period in amphibians.

pronephros
Figures 6.28, 6.38, 6.45, 6.51, 6.55-6.57, 6.62, 6.63
also see nephrotome; pronephric ducts; pronephric tubules
The first and most cranial kidney to form; it derives from buds of nephrotome which hollow out to form tubules–one pair per body segment; one end of each tubule opens as a nephrosome into the coelom; the tubules link together to form the pronephric duct which grows caudally along the somites to the cloaca; it is vestigial in amniotes but large in lower vertebrates and functions in the larval stage; the pronephric duct is appropriated by the mesonephros in amniotes.

pronucleus
Figures 2.4, 2.8
A haploid nucleus found in fertilized eggs, one pronucleus derives from the sperm and a second one from the egg; it may fuse with the other pronucleus present in the fertilized egg, or it may enter prophase of the 1st cleavage division separately.

prophase
Figure 1.3
also see anaphase; metaphase; telophase
A phase in cell division prior to metaphase and before chromosomes condense.

prosencephalon
Figures 6.21, 6.22, 6.26-6.31, 6.33, 6.34, 6.40-6.42, 6.45, 7.30-7.34, 7.45
also see diencephalon; telencephalon
The cranial primary brain vesicle; it forms two lateral evaginations (optic vesicles) and a ventral evagination (the infundibulum); it differentiates into a cranial telencephalon and a caudal diencephalon. *Synonym:* forebrain.

pulmonary arteries
Figures 6.57, 7.85, 8.6, 8.7, 8.22, 8.24
They connect the pulmonary trunk with the lungs; their basal sections derive from aortic arch 6.

pulmonary trunk
Figure 8.20
The trunk of the pulmonary arteries; it connects with the right ventricle and derives from the conotruncus by longitudinal division of the latter.

pulmonary veins
Figures 6.57, 8.24
The vessels carrying blood from the lungs to the left atrium; they arise as outgrowths of the left atrium, or in birds from the sinus venosus, and connect with the pulmonary plexus.

rami cranial nerve V
Figure 8.6
also see mandibular ramus; cranial nerve V; maxillary ramus; cranial nerve V; ophthalmic ramus; cranial nerve V
The main branches of cranial nerve V (trigeminal nerve) consisting of, from cranial to caudal: ophthalmic ramus, maxillary ramus and mandibular ramus.

Rathke's pouch
Figures 7.44, 7.45, 7.47, 7.52, 7.64, 7.67, 7.68, 7.74, 7.86, 8.2,

8.3, 8.8, 8.15
also see hypophysis
A dorsal evagination of the stomodeum extending under the diencephalon to the infundibulum in amniotes; it becomes isolated from the stomodeum and forms the pars distalis (cranial lobe), the pars intermedia (intermediate lobe) and pars tuberalis of the hypophysis.

rays
Figure 3.1
In *C. elegans* adult males, supporting structures of the fan.

rectum
Figures 3.1, 3.21, 7.5, 8.35, 8.36
The caudal segment of the large intestine; it is formed by splitting off from the dorsal side of the cloaca.

renal pelvis of metanephros
Figure 8.37
also see metanephric diverticulum
The expanded distal end of the metanephric diverticulum; it is surrounded by metanephrogenic mesenchyme and it forms the pelvis, calyces and collecting tubules of the metanephros.

residual bodies
Figures 1.2, 7.2
Granules of degenerating cytoplasm sloughed off differentiating spermatids; they are phagocytized by the Sertoli cells.

residual spermatogonia
Figure 6.2
Large reserve germ cells of amphibia; they may proliferate mitotically to replace cells that have matured into sperm.

retina
Figures 6.55, 6.59
also see sensory layer of optic cup; optic cups; pigmented layer of optic cup
The inner sensory and pigmented layers of the eye; it is derived from two layers of the optic cup–its inner wall, which becomes the sensory part of the retina, and its outer wall, which becomes the pigmented epithelium of the retina.

rhombencephalon
Figures 6.22, 6.26, 6.28, 6.29, 6.35-6.37, 6.40, 6.41, 6.48-6.50, 7.31, 7.33, 7.36-7.38
also see metencephalon; myelencephalon
The third caudal primary brain vesicle extending from the mesencephalon to the spinal cord; it divides into an cranial metencephalon and a caudal myelencephalon; it forms the cerebellum, pons and medulla. *Synonym:* hindbrain.

right atrium
Figures 8.6, 8.21-8.24
also see atrium
The right division of the primitive atrium, separated from the left atrium by the atrial septum; it receives blood from the sinus venosus (or later from the superior and inferior vena cavae) and it delivers blood through the interatrial foramen to the left atrium, and through the right atrioventricular canal to the right ventricle; after breathing begins, the interatrial foramen closes.

right diverticulum
Figure 5.4
also see dorsal diverticulum
In *Amphioxus,* it arises from the dorsal diverticulum, extends ventrally and expands to form the thin-walled head cavity.

right horn of sinus venosus
Figure 8.22
also see sinus venosus
The part receiving blood from the right common cardinal, right vitelline and right umbilical veins, and later, from the inferior vena cava; it is eventually incorporated into the right atrium with its veins.

right umbilical vein
see umbilical vein

right ventricle
Figures 8.6, 8.21, 8.22, 8.24, 8.25
also see ventricle
A thick-walled heart chamber formed by the partitioning of the primitive ventricle by the ventricular septum; it receives blood from the right atrium and it delivers blood to the pulmonary trunk and ductus arteriosus.

right vitelline artery
Figure 7.46
also see vitelline arteries
The arterial supply for the right half of the yolk sac.

right vitelline vein
Figure 7.46
also see vitelline veins
The venous return for the right half of the yolk sac.

root of cranial nerve V
Figure 7.71, 8.11
also see cranial nerve V
The part of cranial nerve V connecting the semilunar ganglion to the metencephalon.

root of cranial nerve VII
Figure 8.12
also see cranial nerve VII
That part of cranial nerve VII connecting the geniculate ganglion to the myelencephalon.

sclerotome
Figures 6.51, 7.59, 7.75, 8.17, 8.20, 8.31, 8.39, 8.40
also see somites
The medial, mesenchymal division of the somite; it arises from cells of the medioventral wall of the early somite and envelopes the notochord and spinal cord; sclerotomes split transversely, and adjacent halves fuse to form the rudiments of the vertebrae and ribs.

secondary mesenchyme
Figures 4.12-4.14
also see primary mesenchyme
Cells that migrate into the blastocoel from the wall of the archenteron during gastrulation in sea urchins; it occupies the animal part of the blastocoel, forming skeleton and muscle.

secondary spermatocytes
Figures 1.2, 1.3, 1.10, 1.13, 7.3
Male germ cells formed from primary spermatocytes by the 1st maturation division; they undergo at once the 2nd maturation division to form spermatids; they are distinguished from both primary spermatocytes and spermatids by their intermediate size.

Seessel's pocket
see preoral gut

segmental arteries
Figures 7.85, 8.7
A series of small branches of the aorta arising between the somites; at cervical regions they contribute to the subclavian and vertebral arteries.

segmental plate mesoderm
Figures 7.20, 7.27, 7.30, 7.31, 7.40, 7.41, 7.43, 7.62
Paraxial mesoderm extending caudally from the last somite; it segments to form somites.

segmental cavity
see blastocoel

semilunar ganglion
Figures 8.5, 8.6, 8.11-8.14
also see ganglion of cranial nerve V

seminiferous tubules
Figures 1.1, 6.1, 7.1
Tubules within the testis; they are bounded by a thin basement membrane of connective tissue and contain the male germ cells and Sertoli cells.

sensory layer of optic cup
Figures 6.47, 6.59, 7.75, 8.17
also see optic cup
The inner layer of the optic cup; it arises from the lateral wall of the optic vesicle and forms the sensory layer of the retina and the optic nerve fibers.

sensory root
see dorsal root of spinal nerve

septum
Figure 6.1
In the frog testis, connective tissue membranes enclosing the seminiferous tubules.

septum spurium
Figure 8.23
A temporary ridge on the dorsal wall of the right atrium extending to the valve of the sinus venosus.

serosa
see chorion

Sertoli cells
Figures 1.2, 1.5, 1.6, 6.2, 7.2
The sperm nurse cells of vertebrates; in mammals, tall, columnar, phagocytic cells extending from the basement membrane to the lumen of the seminiferous tubule; the outline of these cells is irregular and obscure; their nucleus is light staining with a prominent nucleolus; differentiating germ cells become embedded in cytoplasmic pockets in Sertoli cells and withdraw at maturity; they form part of the blood-testis barrier. *Synonym:* sustentacular cells.

shell
Figures 7.6, 7.48, 7.89
also see egg shell
The outer wall of the egg; in birds, it is composed of calcium carbonate crystals impregnating protein fibers; it is perforated by numerous microscopic pores to permit respiration; the pores are sealed by the cuticle composed of dry albumen; it is reinforced by the shell membranes attached to its inner surface; the shell is secreted around the egg while it passes through the uterus.

shell membranes
Figures 7.6, 7.48, 7.89
In the bird egg, a pair of flexible membranes composed of protein fibers and attached to the shell which they reinforce; at the blunt end of the egg the membranes separate to enclose the air chamber.

sinoatrial region
Figures 7.30-7.33
The most caudal division of the heart; it forms the right and left atria and the sinus venosus.

sinus venosus
Figures 6.32, 6.44, 6.57, 6.61, 7.43-7.47, 7.57, 7.63, 7.64, 7.66-7.69, 7.75, 7.84-7.88, 8.22, 8.24, 8.25, 8.29
also see left horn of sinus venosus; right horn of sinus venosus; transverse sinus venosus
Initially the most caudal chamber of the heart, receiving the venous return and delivering it to the atrium; after the partitioning of the atrium, it empties into the right atrium; it disappears as a heart chamber by atrophy and by incorporation into the atria; it originates the heart beat and later transfers that function to the atrium by forming the sinoatrial node.

skeleton
Figure 4.19
In sea urchin embryos and larvae, calcite spicules formed by mesenchymal cells; it supports the arms of the pluteus larvae.

small intestine
Figures 8.6, 8.35, 8.36
also see duodenum; intestinal loop; intestine
The segment of gut after the stomach; it arises from the foregut and midgut in amphibians, and from the foregut and hindgut in amniotes.

smooth muscle
Figure 9.1
also see myometrium
Nonstriated involuntary muscle located in the walls of hollow organs as in the uterus; its contraction is controlled by hormones (oxytocin, adrenalin and noradrenalin) and the autonomic nervous system.

somatic gonad cells
Figure 3.2
In *C. elegans* embryos, some of the cells derived from the MS blastomere; they give rise to gonad tissue.

somatic mesoderm
Figures 6.38, 7.26, 7.33, 7.35-7.37, 7.40-7.42, 7.79
also see lateral plate mesoderm; somatopleure
The cellular layer immediately outside and originally dorsal to the coelom; it arises by splitting of the lateral plate mesoderm; it forms parietal peritoneum and, by fusion with myotomes, dermatomes and epidermis, it forms the body wall and limbs; in the extraembryonic area, it fuses with the ectoderm to form the somatopleure of the amnion and chorion.

somatopleure
Figures 7.25, 7.26, 7.29, 7.35-7.38, 7.42, 7.45, 7.48
also see amnion; chorion
A double-layered membrane composed of ectoderm and somatic mesoderm; it contributes to the body wall and extends into the extraembryonic area; it forms the amniotic folds which by enveloping the embryo, transform the extraembryonic

somatopleure into amnion and chorion.

somites
Figures 5.4, 5.15, 5.20, 6.16, 6.17, 6.20, 6.24, 6.26, 6.28-6.30, 6.37-6.43, 6.50-6.54, 6.65, 7.19, 7.20, 7.26, 7.30-7.33, 7.39, 7.43-7.45, 7.47, 7.55-7.58, 7.61, 7.63, 7.64, 7.67-7.69, 7.71-7.75, 7.78, 7.80-7.84, 7.86, 7.87, 8.2-8.5, 8.12, 8.15, 8.30
also see dermatome; myotome; sclerotome
The segments of the paraxial mesoderm; they form first at the caudal end of the myelencephalon and extend progressively as a series of paired blocks caudally into the tail; they are separated by intersomitic grooves and attach laterally to the nephrotomes; they are primary segments of the body which establish all other segmental patterns; they differentiate into a lateral dermatome, middle myotome, and medial sclerotome. *Synonym:* epimere.

sperm
Figures 1.3, 1.5, 1.10, 2.1-2.3, 2.7, 2.8, 5.2, 6.1, 6.3, 7.2
also see differentiating spermatid; spermiogenesis
The mature male germ cell; in vertebrates, a small, haploid, highly specialized, flagellated cell which can attach to and penetrate egg membranes to activate the egg; it forms from a spermatid through a complex differentiation called spermiogenesis. *Synonym:* spermatozoon.

sperm aster
Figures 4.2, 4.3
Fibers radiating from the sperm's centrioles within the egg cytoplasm after fertilization; together with the spindle, it forms the mitotic apparatus of the fertilized egg (zygote).

spermatid
Figures 1.2, 1.3, 1.5, 1.6, 1.10, 1.13-1.15, 6.1-6.3, 7.2, 7.3
also see differentiating spermatid
A small haploid germ cell of the testis; it forms from a secondary spermatocyte during the 2nd maturation division; it is embedded in a pocket within the Sertoli cell, and it differentiates into a sperm.

spermatocytes
Figures 6.1, 6.3
also see primary spermatocytes; secondary spermatocytes
Cells formed in males during spermatogenesis.

spermatogonia
Figures 1.2, 1.5, 1.6, 1.10-1.12, 7.2
The "stem" germ cells of the testis; each divides mitotically within the germinal epithelium; they are located near the basement membrane of the seminiferous tubule outside the blood-testis barrier; they may enter a prolonged growth phase forming primary spermatocytes.

sperm heads
Figures 1.2, 1.4, 4.2, 6.2
also see sperm; sperm asters; sperm midpieces; sperm pronucleus; sperm tails
That part of the sperm containing the nucleus and acrosome.

spermiogenesis
Figures 6.3, 7.3
also see differentiating spermatid
The final phase of spermatogenesis during which the spermatid transforms or differentiates into a sperm; during this period the cells are enveloped by the cytoplasm of the Sertoli cells which probably provides the special environment required for the transformation.

sperm midpieces
Figure 1.4
also see sperm; sperm asters; sperm heads; sperm pronucleus; sperm tails
The portion of the sperm containing mitochondria; mitochondria within the midpiece are used for the production of energy for sperm motility.

sperm pronucleus
Figure 4.3
also see sperm; sperm asters; sperm heads; sperm midpieces; sperm tails
The nucleus of the sperm within the cytoplasm of the egg shortly after fertilization.

sperm tails
Figures 1.1, 1.4
also see centrioles; differentiating spermatid; immature sperm; sperm; sperm asters; sperm heads; sperm midpieces; sperm pronucleus
The flagellae of the sperm; they are derived from the cytoplasm of spermatids; they are each composed of an axial filament arising from centrioles near the head, a mitochondrial sheath, fibers and a plasma membrane.

spicules
Figure 3.1
In *C. elegans* adult males, supporting structures within the tail, near the cloaca.

spinal cord
Figures 6.17, 6.24-6.26, 6.28-6.30, 6.38-6.42, 6.45, 6.51-6.57, 6.64, 6.65, 7.30, 7.39, 7.40, 7.43, 7.44, 7.47, 7.57-7.59, 7.61-7.64, 7.67, 7.69, 7.72, 7.74, 7.77, 7.78, 7.80-7.83, 7.86, 7.88, 8.3, 8.8, 8.9, 8.15, 8.17-8.20, 8.24, 8.26, 8.28, 8.30, 8.32, 8.35, 8.36, 8.38-8.40
also see neural tube
The central nervous system caudal to the brain; it derives from the caudal neural tube and bears a pair of spinal nerves for each body segment; its wall differentiates into an inner ventricular layer, a middle intermediate layer and an outer marginal layer; the latter two layers are rudiments of gray matter and white matter, respectively.

spinal ganglia
Figures 6.45, 6.52, 6.54, 6.64, 6.65, 7.57, 7.72, 7.73, 7.76, 8.3, 8.6, 8.8, 8.18, 8.20, 8.24, 8.26, 8.28, 8.30, 8.37, 8.39, 8.40
Ganglia borne on dorsal roots of spinal nerves; they derive from neural crest and supply sensory nerve fiber of the spinal nerve.
Synonym: dorsal root ganglia.

spinal nerves
Figures 8.28, 8.30, 8.39, 8.40
also see dorsal root of spinal nerve; spinal ganglia; ventral root of spinal nerve
Paired nerves emerging from the spinal cord at each body segment; each is connected to the spinal cord by a dorsal root bearing a spinal ganglion and by a ventral root; the spinal nerve trunk divides immediately into a dorsal ramus and a ventral ramus; a ramus communicans connects to autonomic ganglia.

splanchnic mesoderm
Figures 6.38, 7.24, 7.29, 7.33, 7.35, 7.37, 7.41, 7.42, 7.48, 7.79
also see lateral plate mesoderm; splanchnopleure
The cell layer between the coelom and endoderm; it arises by splitting of the lateral plate mesoderm; it fuses with the endoderm to form the wall of the gut and respiratory tract; it forms mesenteries, visceral peritoneum, heart and germinal epithelium;

in the extraembryonic area, it fuses with the endoderm to form the splanchnopleure of the yolk sac and allantois.

splanchnopleure
Figures 7.24, 7.29, 7.32, 7.34-7.38, 7.42, 7.45, 7.70
also see allantois; yolk sac
A double membrane composed of splanchnic mesoderm and endoderm; it forms the gut wall and extends into the extraembryonic area to form the yolk sac and allantois.

stem placental villi
Figure 9.2
also see placental villi
In the placenta, the trunk or large branch of each placental villus; they contain arteries and veins with fetal blood which arises from the umbilical arteries and veins; they supports terminal villi.

stomach
Figures 4.15-4.19, 4.31-5.1, 6.55, 6.57, 6.62, 6.63, 7.76, 8.3, 8.6, 8.28, 8.31, 8.33
An enlarged segment of the foregut caudal to the esophagus; it derives from the archenteron in the sea urchin, starfish and *Amphioxus;* its lining epithelium and glands form the gut endoderm, but its muscle, blood vessels and connective tissue develop from the splanchnic mesoderm; in birds, the stomach differentiates into a proventriculus and a gizzard.

stomodeum
Figures 4.14, 4.30, 6.17-6.21, 6.26, 6.28, 6.29, 6.32, 6.34, 6.40, 6.41, 6.44, 6.45, 6.47, 7.33, 7.35, 7.43-7.45, 7.47, 7.53, 7.64, 7.67-7.70, 7.73, 7.74, 8.3, 8.4, 8.17-8.19
also see mouth
The ectodermal rudiment of the mouth; an invagination in the cranioventral ectoderm of the head that contacts the cranial wall of the foregut; its floor is the oral membrane; rupture of the membrane opens the mouth into the pharynx; a rudiment of the hypophysis, called Rathke's pouch, evaginates from the dorsal wall of the stomodeum.

stratum granulosa
Figures 1.7, 1.8, 7.4
also see cumulus oophorus
The inner stratified epithelium of large ovarian follicles; it derives from the follicle cells or primary follicles; in mammals, it contributes to the corpus luteum after ovulation.

stroma
Figure 1.7, 9.1
also see connective tissue; endometrium
The connective tissue framework of an organ; in the mammalian ovary and endometrium, it consists of a dense population of elongated cells and some delicate fibers.

structural, other cells
Figure 9.2
In *C. elegans* embryos, some of the cells derived from the AB and MS blastomeres; they give rise to a variety of tissue.

subcardinal anastomosis
Figure 8.38
also see subcardinal veins
A medial interconnection between the right and left subcardinal veins; it contributes to the pre-renal segment of the inferior vena cava. *Synonym:* subcardinal sinus.

subcardinal veins
Figures 7.85, 8.7, 8.35-8.37

Primitive paired veins of the trunk; they lie ventral to the mesonephroi and parallel to the caudal cardinal veins which they mostly replace; they subsequently contribute to the inferior vena cava and its branches.

subcephalic pocket
Figures 7.23, 7.34
also see head fold of the body
A cavity beneath the embryonic head formed by the head fold of the body as it pushes under the head; it is lined by the ventral head ectoderm of the somatopleure; it lengthens as the head grows forward and the head fold is drawn caudally.

subclavian arteries
Figures 7.85, 8.6, 8.7
The arteries of the shoulder and forelimb; they arise by the enlargement of the 7th intersegmental arteries; in mammals, the right subclavian also receives contributions from the right aortic arch 4 and the right dorsal aorta.

subclavian veins
Figures 7.85, 8.7, 8.22
The veins of the forelimbs; they connect at first to caudal cardinal veins but later they shift to the cranial cardinal veins.

subgerminal cavity
Figures 7.9, 7.10, 7.15, 7.23
A space beneath the hypoblast of the area pellucida in birds.

subintestinal vein
Figure 7.85
A vein in pig embryos extending from the base of the tail along the ventral margin of the intestine to the vitelline veins; initially it drains the allantois, caudal limb buds and intestine; it is mostly replaced by the development of the allantoic veins.

subnotochordal rod
Figures 6.29, 6.39, 6.42
A strand of cells lying between the midgut and notochord in amphibians; it is of endodermal origin; it degenerates.

superior ganglion
Figures 8.2, 8.6, 8.11
also see ganglion of cranial nerve IX
The dorsal ganglion of cranial nerve IX; with the petrosal ganglion, it supplies the sensory fibers to the nerve.

superior mesenteric artery
Figures 8.32, 8.35-8.37
also see vitelline arteries
The arterial supply of the small intestine; it is derived from the vitelline arteries.

superior mesenteric vein
Figure 8.35
The main branch of the portal vein; it drains the digestive tract.

superior vena cava
Figure 8.29
The main venous trunk vessel draining the head, neck and forelimb regions directly into the right atrium; it is derived from the right common cardinal and right cranial cardinal veins.

sustentacular cells
see Sertoli cells

sympathetic ganglia
Figures 8.22, 8.30
A series of paired ganglia dorsal to the aorta and connected to the spinal nerves by the rami communicans; they are part of the autonomic nervous system; they derive from the neural crest.

syntrophoblast
Figure 9.4
also see villus syntrophoblast
The outer epithelial covering of placental villi; the part of the chorionic surface and placental barrier in contact with maternal blood; a true syncytium; possesses a brush border on its free surface; it is partly underlain by and derived from the cytotrophoblast; it is the probable site of the synthesis of placental hormones.

tail
Figures 3.22, 3.23, 7.63, 7.67, 8.3, 8.4, 8.9, 8.31
also see posterior; tail bud
The extension of the body caudal to the anus; it derives from the tail bud.

tail bud
Figures 6.26, 6.27, 6.29, 7.43, 7.47, 7.67, 7.68, 7.84, 8.1
The rudiment of the tail and caudal trunk; a mass of undifferentiated tissue projecting from the caudal end of the embryo; it derives from the primitive streak in amniotes; it contributes to the neural tube and somites.

tail end
see posterior

tail fin
Figures 6.30, 6.31, 6.40-6.44, 6.55, 6.56, 6.65
also see caudal fin
A bladelike extension of the border of the tail in amphibians and *Amphioxus;* it is continuous cranially with the dorsal fin.

tail fold of the body
Figures 7.43, 7.62, 7.82
A depressed fold encircling the tail bud and connecting cranially with the body folds; it forms part of the boundary between the embryonic and extraembryonic areas; it undercuts the tail bud and caudal trunk, forming the hindgut.

tail flexure
Figure 7.84
also see cervical flexure; cranial flexure
One of several ventral bends in the body axis giving the amniote embryo a compact C-configuration; its forms in the region of the tail.

tail gut
see postanal gut

teeth
Figures 6.56-6.58
In tadpoles 3 or 4 rows of horny epidermal papillae attached to the jaw cartilages; they are frequently shed and replaced; they are lost during metamorphosis and replaced by true teeth.

telencephalon
Figures 6.46, 6.56, 6.58, 7.43, 7.44, 7.47, 7.55, 7.56, 7.63, 7.64, 7.67-7.70, 7.76, 7.77, 7.86, 7.87, 8.1, 8.3, 8.8, 8.18-8.20
also see cerebral hemispheres
The cranial division of the prosencephalon; its greatly enlarged

roof forms the cerebral hemispheres; its floor forms the olfactory bulbs, hippocampus and corpus striatum; its cavities are the lateral ventricles of the brain.

telophase
see anaphase; metaphase; prophase
A phase in cell division in which chromosomes have moved along spindle fibers to the spindle poles (centrioles).

testicular cyst
Figure 1.10
also see testicular lobe wall
In the grasshopper testis, a compartment within a testicular lobe bounded by connective tissue septa and containing a group of germ cells at the same stage of maturation.

testicular lobe wall
Figures 1.10, 1.15
also see testicular cyst
In the grasshopper testis, the connective tissue capsule enclosing a lobe or major division in the testis; the lobe is divided into cysts, each containing a cluster of germ cells; the apical end of the lobe contains proliferating spermatogonia, with more mature germ cells extending toward the opposite end which opens into a vas deferens.

testis
Figures 1.1, 1.2, 1.5, 1.6, 1.10, 1.12-1.15, 3.1, 6.1, 6.2, 7.1, 7.2
The male gonad; the organ in which sperm differentiate; it secretes the male sex hormone testosterone in vertebrates.

tetrad
Figure 2.7
see bivalent

theca externa
Figures 1.8, 6.4
The outer connective tissue layer of Graafian follicles; it arises from the stroma of the ovary; it is the outer wall of the ovary in amphibians.

theca folliculi
Figures 1.7, 7.4
also see theca externa; theca interna
The outer capsule of ovarian follicles; these layers form from the connective tissue stroma as follicles grow; in mature follicles, the theca folliculi differentiate into the theca interna and theca externa.

theca interna
Figures 1.7, 1.8, 6.4
A vascular layer between the theca externa and the stratum granulosa of large ovarian follicles; it contains endocrine gland cells, connective tissue and blood vessels; it contributes to the corpus luteum after ovulation or to the interstitial tissue after follicular atresia in mammals.

third ventricle
Figures 7.47, 7.68, 7.86, 8.8
also see lateral ventricles
Originally the enlarged neural canal of the telencephalon; it later divides into the lateral ventricles of the cerebral hemispheres and the definitive third ventricle of the diencephalon; its thin roof forms a choroid plexus.

thyroid (rudiment) gland
Figures 6.29, 6.35, 6.41, 6.45, 6.48, 6.55, 6.56, 6.59, 7.47, 7.54, 7.55, 7.67, 7.68, 7.73, 7.86, 8.3, 8.17
An endocrine gland in the throat region; its forms as a ventral diverticulum of the pharynx at the level of the of the second branchial arches; the rudiment bifurcates and migrates caudally, becoming isolated from the pharynx.

tongue
Figures 8.5, 8.8, 8.9, 8.16
also see copula; lateral lingual swellings; tuberculum impar
In mammals, arises by fusion of several elevations on the floor of the mouth and pharynx; these elevations include two lateral lingual swellings and a median tuberculum impar on the mandible, the copula on branchial arch 2 and contributions from branchial arches 3 and 4; later the embryonic tongue is invaded by muscle from a more caudal level and is innervated by cranial nerve XII.

trabeculae carneae
Figure 8.23
Interlacing muscle bands in the wall of the ventricle of the heart; they contribute to the formation of the ventricular septum.

trachea
Figures 6.56, 7.86, 8.6, 8.8, 8.21, 8.22, 8.27
The part of the respiratory tract connecting the laryngotracheal groove with the lung buds; it arises with the lung buds as a ventrocaudal diverticulum of the pharynx; its muscle and connective tissue develop from the splanchnic mesoderm of the ventral mesoesophagus.

transverse septum
Figures 6.45, 6.57, 7.58, 8.25, 8.26
A mass of mesenchyme caudal to the heart, incompletely separating the pericardial cavity from the peritoneal cavity; it encloses the veins that enter the heart; the liver is attached to its caudal face; it contributes to the diaphragm in mammals.

transverse sinus venosus
Figure 8.25
also see sinus venosus
A narrow middle part carrying blood from the left horn to the sinoatrial opening; it eventually forms the coronary sinus.

trigeminal nerve
see cranial nerve V

trigeminal placode
Figure 7.51
also see ganglion of cranial nerve V; neural crest
A thickened plate of head ectoderm dorsal to the mandibular arch; cells detach from its under surface and join the neural crest cells, forming the semilunar ganglion of cranial nerve V.

trophoblast
Figure 9.4
also see villus cytotrophoblast; villus syntrophoblast
The outer epithelial covering of the chorion and placental villi; the principal component of the placental membrane; it probably secretes the placental hormones.

truncus arteriosus
see ventral aorta

tuberculum impar
Figure 8.15
also see tongue
A median elevation of the mandible in the floor of the mouth; it fuses with the lateral lingual swellings to form the body of the

tongue.

tunica albuginea
Figures 1.1, 1.7, 1.8, 6.2
A fibrous connective tissue capsule or membrane enveloping the ovary and testis.

ultimobranchial body
Figure 8.19
An evagination from the caudal surface of each fourth pharyngeal pouch; each represents pharyngeal pouch 5; it fuses with the thyroid rudiment, forming the parafollicular cells of the thyroid gland. *Synonym:* postbranchial body.

umbilical arteries
Figures 7.85, 7.89, 8.6-8.9, 8.32, 8.35-8.37
The arterial blood supply to the chorioallantois of birds and the placenta of mammals; a pair of vessels arising from the caudal end of the aorta; it forms the common iliac and hypogastric arteries in mammals, and after birth, the lateral umbilical ligaments. *Synonym:* allantoic arteries.

umbilical cord
Figures 8.4, 8.5, 8.31, 8.32
The narrowed connection of the embryo to the extraembryonic membranes; its outer wall is amnion and may contain the yolk stalk, allantoic stalk, vitelline blood vessels, umbilical blood vessels and a gelatinous connective tissue; in birds, it separates from the umbilicus just before hatching; in mammals, it is bitten in two after birth, the stump dropping off in a few days.

umbilical veins
Figures 7.76, 7.79, 7.85, 7.88, 8.2, 8.6-8.9, 8.29, 8.31, 8.32, 8.35-8.37
Initially, paired embryonic vessels draining the allantois of birds or the placenta of mammals; the left vein atrophies early, the right atrophies after hatching or birth, forming, in mammals, the ligamentum teres. *Synonym:* allantoic veins.

ureter
Figures 8.6, 8.37
The excretory duct of the metanephros; it derives from the stalk of the metanephric diverticulum, connecting at first with the mesonephric duct, its site of origin; it later shifts to the cloaca in birds or to the urinary bladder in mammals. *Synonym:* metanephric duct.

ureteric bud
see metanephric diverticulum

urogenital sinus
Figures 8.5, 8.6, 8.9, 8.36
also see cloaca
A chamber split from the ventral part of the cloaca of mammals; it receives the mesonephric ducts, Mullerian ducts and allantoic stalk; it contributes to the bladder and forms the urethra; in females, it forms the vestibule of the vagina as well.

uterine cavity
Figure 9.1
also see endometrium; uterine epithelium; uterine glands
The lumen of the uterus; it is lined by the uterine epithelium of the endometrium.

uterine epithelium
Figure 9.1

also see endometrium; uterine epithelium; uterine glands
A simple columnar glandular epithelium with some ciliated cells lining the uterine cavity; part of the endometrium.

uterine glands
Figures 9.1, 9.4
also see endometrium
Tubular glands of the endometrium; they are active during the secretory phase of the menstrual cycle, releasing a nutrient mixture of glycogen, mucinogen and fat which provides a supportive medium for the developing embryo prior to implantation.

uterus
Figure 3.1, 7.5
In birds, the large terminal segment of the oviduct; it forms the egg shell as the egg is held there; it passes the finished egg into the cloaca at laying; in viviparous mammals, reptiles and some fishes, it provides metabolic support for the developing young.

vagus nerve
see cranial nerve X

valve of sinus venosus
Figures 8.22, 8.23
Valve of the sinoatrial opening into the right atrium.

vas deferens
Figure 3.1
A portion of the male reproductive system. It transports sperm from the testes.

vegetal hemisphere
Figures 4.9, 5.13
also see animal hemisphere; vegetal pole
That half of the egg (oocyte) or early embryo containing the most yolk, the other half being the animal hemisphere; the vegetal pole lies at its center and opposite the animal pole.

vegetal pole
Figures 4.26, 4.27, 6.10, 6.17
also see vegetal hemisphere
The end of the embryonic axis centered in the yolky region of the egg; it is opposite the animal pole.

veins
Figure 7.1, 9.4
In the testis, veins, arteries and capillaries branch in the interstitial connective tissue between the seminiferous tubules; these blood vessels do not penetrate the tubules; many veins are also present in the uterus.

velar plate
Figure 6.60
Laterally projecting plates on the floor of the pharynx of tadpoles which reduce the openings of the pharynx into the opercular chamber to narrow slits.

ventral
Figures 3.12, 3.17-3.19
The side of the embryo at which the gut forms.

ventral aortae
Figures 7.28, 7.32, 7.35
also see aortic sac
The outlets of the embryonic heart; they lie below the floor of the

pharynx and conduct blood from the conotruncus to the aortic arches; they fuse to from the aortic sac.

ventral cleft
Figure 3.15
In *C. elegans* embryos, a ventral indentation that appears as the embryo begins to elongate into a worm.

ventral ectoderm
Figure 5.15
also see ectoderm; epidermis
The outer or ectodermal layer covering the ventral surface of the embryo; it forms the epidermis in later development.

ventral lip
Figures 5.3, 5.14, 5.19, 6.12-6.16
also see dorsal lip; lateral lips
The margin of the blastopore toward the animal pole and at the ventral side of the embryo; it derives from the ventral marginal zone and forms ventral mesoderm in amphibians.

ventral liver bud
see caudal liver diverticulum

ventral mesentery
Figures 7.58, 7.76, 8.34
also see dorsal mesocardium; lesser omentum; ventral mesoesophagus
A double layer of splanchnic mesoderm attaching parts of the foregut to the ventral body wall; it forms the transient mesocardia, roots of the lungs, lesser omentum and falciform ligament of the liver.

ventral mesocardium
Figure 7.29
also see dorsal mesocardium
A temporary mesentery attaching the ventral wall of the heart to the body wall.

ventral mesoderm
Figures 6.16, 6.29
also see lateral plate mesoderm
The extension of the lateral plate into the ventral body region; it is split by the coelom into the somatic and splanchnic mesoderm.

ventral mesoesophagus
Figure 8.25
also see dorsal mesocardium
The ventral mesentery of the esophagus; it forms the roots of the lungs and the transient mesocardia.

ventral pancreatic rudiment
Figures 8.6, 8.32-8.34
also see dorsal pancreatic rudiment
A ventral evagination of the liver which grows, branches and fuses with the dorsal pancreatic rudiment to form one glandular mass of the adult pancreas.

ventral ramus of spinal nerve
Figures 8.28, 8.30
also see spinal nerve
The main ventral branch of the spinal nerve trunk; each innervates the viscera, body wall and limbs.

ventral root of spinal nerve
Figures 8.28, 8.30, 8.39, 8.40
also see spinal nerve

The ventral division of a spinal nerve connecting the trunk of the nerve to the basal plate of the spinal cord; it is composed of motor nerve fibers arising from neuroblasts in the intermediate layer of the basal plate. *Synonym:* motor root.

ventral veins
Figure 8.7
Transient veins extending along the ventral margin of the mesonephros in pig embryos and draining into the caudal cardinal veins; they are subsequently replaced by the subcardinal veins.

ventricle
Figures 6.32, 6.44, 6.50, 6.57, 6.60, 7.30, 7.32, 7.33, 7.37, 7.43-7.47, 7.57-7.59, 7.63, 7.64, 7.66-7.70, 7.76, 7.77, 7.84-7.87, 8.3, 8.5, 8.8, 8.9, 8.21, 8.22, 8.24, 8.25, 8.28
also see heart; left ventricle; right ventricle
The thick-walled heart chamber that, in the embryo, receives blood from the atrium and delivers it under high pressure to the conotruncus; in amniotes, it is partitioned into the right and left ventricles, delivering blood to the pulmonary trunk and ascending aorta, respectively.

ventricular layer
Figures 8.28, 8.30
The inner layer of the primitive neuroepithelial cells of the neural tube; by proliferation of cells, it supplies neuroblasts for the intermediate and marginal layers, and subsequently forms the ependymal of the spinal cord and brain.

vertebral arteries
Figures 7.85, 8.7, 8.15
A pair of longitudinal vessels extending cranially from the subclavian arteries to the basilar artery under the myelencephalon; with the internal carotids, it provides the arterial supply to the brain; each arises from an anastomosis of cranial intersegmental arteries.

vesicular follicle
see Graafian follicle

villus cytotrophoblast
Figure 9.3
also see placental villi
The inner epithelial layer covering placental villi; its cells are well defined and may show mitotic figures; it probably forms the overlying syntrophoblast. *Synonym:* Langerhans cells.

villus syntrophoblast
Figures 9.2, 9.3
also see peripheral syntrophoblast; placental barrier; villus cytotrophoblast
The outer epithelial covering of placental villi; the part of the chorionic surface and placental barrier in contact with maternal blood; a true syncytium; possesses a brush border on its free surface, it is partly underlain by and derived from the cytotrophoblast; it is the probable site of the synthesis of placental hormones.

visceral cleft
see branchial cleft

visceral groove
see branchial groove

visceral pouch
see pharyngeal pouch

vitelline arteries
Figures 6.45, 7.43, 7.45, 7.46, 7.61, 7.66-7.68, 7.70, 7.80, 7.84-7.86, 7.89, 8.6, 8.7
The arterial supply of the yolk sac; they arise as ventral branches of the dorsal aortae; they form the superior mesenteric artery and, in mammals, the celiac and inferior mesenteric arteries as well.

vitelline membrane
Figures 2.1, 3.3-3.5, 3.26, 3.20, 4.4, 4.20, 7.6, 7.48
A membrane enveloping the egg or oocyte; it lies immediately outside the plasmalemma; it is formed while the oocyte is in the ovary and, in some species, after fertilization, it separates from the egg to form the fertilization membrane.

vitelline plexus
Figures 7.30, 7.39, 7.41, 7.46
A network of small vessels in the yolk sac; some enlarge to form the vitelline veins and arteries.

vitelline veins
Figures 6.32, 6.43-6.45, 6.57, 7.28, 7.30, 7.32, 7.33, 7.38, 7.45, 7.46, 7.58, 7.59, 7.66, 7.67, 7.70, 7.77-7.79, 7.84-7.86, 7.88, 7.89, 8.7, 8.29
also see cranial vitelline veins; caudal vitelline vein
Vessels that provide initially the venous return from the yolk sac; their proximal ends fuse, forming the ductus venosus; they also form the hepatic veins, hepatic sinusoids and the portal vein; their distal branches degenerate with the yolk sac; in amphibians, they form around the yolk endoderm of the midgut.

vitelline vessels
Figures 7.40, 7.49, 7.52, 7.63, 7.71, 7.81, 7.89
also see vitelline arteries; vitelline plexus; vitelline veins
The blood vessels of the yolk sac; they arise from the blood islands.

vulva
Figure 3.1
A portion of the female reproductive system. In *C. elegans*, embryos are shed through the vulva to the outside.

wheel organ
Figure 5.1
In *Amphioxus,* a dark-staining ring of ciliated epithelium caudal to the oral hood; it transports mucus with food toward the mouth and into the pharynx.

wing bud
Figures 7.63, 7.67-7.70, 7.78, 7.79, 7.84, 7.86, 7.88
The rudiment of the wing; it arises as a thickening of the somatic mesoderm of the body wall; it later bears an ectodermal thickening, the apical ectodermal ridge; it is homologous to the foreleg bud and arm bud.

Wolffian body
see mesonephros

Wolffian duct
see mesonephric duct

X-chromosome
Figures 1.11, 1.12, 1.13
also see leptotene stage
The sex chromosome that usually occurs double in females and single in males, where it may be associated with a Y-chromosome; it often exists in a contracted or heteropyknotic state.

yolk and yolk granules
Figures 3.15, 3.16, 4.1, 4.4, 4.5, 4.20, 7.4, 7.6, 7.9, 7.10, 7.22, 7.28
A reserve food mixture within the ovum; in birds they form yolk spheres up to 100 microns in diameter, which are composed mainly of lipids and proteins.

yolk endoderm
Figure 6.38
A mass of large yolky cells in the floor of the midgut in amphibians; it derives from the vegetal hemisphere; subsequently its cells disintegrate and the yolk is absorbed.

yolk plug
Figures 6.12-6.16
A mass of large yolky cells filling the blastopore of the amphibian gastrula; it derives from the vegetal hemisphere of the blastula; it invaginates to form the yolk endoderm of the neurula.

yolk sac
Figures 7.13, 7.34-7.39, 7.47-7.50, 7.52, 7.55-7.57, 7.59, 7.60, 7.68-7.71, 7.75-7.78, 7.80-7.83, 7.87-8.3, 8.5, 8.8, 8.29
also see splanchnopleure
A baglike extraembryonic membrane formed as an extension of the midgut; in vertebrates with large eggs, it encloses and absorbs the yolk; in mammals, it is filled with fluid; it arises from splanchnopleure and contains vitelline blood vessels; it is the earliest blood-forming organ and the source of the primordial germ cells; it forms a placenta in some elasmobranchs and mammals (pig).

yolk stalk
Figures 7.48, 8.6
The narrow connection of the yolk sac to the midgut; it contains the vitelline arteries and veins.

Z2 blastomere
see founder cells

Z3 blastomere
see founder cells

zona pellucida
Figures 1.7, 1.8
A thick membrane containing mucopolysaccharide surrounding the eggs of mammals; it is called a zona radiata in the ovary when perforated by cytoplasmic processes of the oocyte and follicle cells; it is penetrated by sperm during fertilization, and it encloses the embryo during cleavage.

zygotene stage
Figures 1.3, 7.3
also see leptotene stage; pachytene stage
The stage of prophase of the first meiotic division when homologous chromosomes pair or synapse; it is preceded by the leptotene stage and it is followed by the pachytene stage.